China's Global Vision for Ecological Civilization

Jiahua Pan

China's Global Vision for Ecological Civilization

Theoretical Construction and Practical Research on Building Ecological Civilization

中国社会科学出版社
CHINA SOCIAL SCIENCES PRESS

Springer

Jiahua Pan
Chinese Academy of Social Sciences
Beijing, China

Translated by
Dan Wu
Xi'an International Studies University, Xi'an,
Shaanxi, China

ISBN 978-981-16-4533-4 ISBN 978-981-16-4534-1 (eBook)
https://doi.org/10.1007/978-981-16-4534-1

Jointly published with China Social Sciences Press
The print edition is not for sale in China (Mainland). Customers from China (Mainland) please order the print book from: China Social Sciences Press.

Translation from the Chinese Simplified language edition: *生态文明建设的理论构建与实践探索* by Jiahua Pan, © China Social Sciences Press 2019. Published by China Social Sciences Press. All Rights Reserved.

This Springer imprint is published by the registered company Springer Nature Singapore Pte Ltd.
The registered company address is: 152 Beach Road, #21-01/04 Gateway East, Singapore 189721, Singapore

Preface

Building an ecological civilization is vital to sustain the Chinese nation's development. In the report delivered at the 19th CPC National Congress, Xi Jinping stressed that we must realize that lucid waters and lush mountains are invaluable assets and act on this understanding. We must continue the Beautiful China initiative to create sound working and living environments for our people and play our part in ensuring global ecological security. The systematic thoughts in building an ecological civilization have contributed Chinese wisdom and plan to solving human problems and represent an important part of Xi Jinping Thought on Socialism with Chinese Characteristics for a New Era.

We need to thoroughly study and implement the spirit of building an ecological civilization in the report of the 19th CPC National Congress, promote the Marxist holistic study of Xi Jinping Thought on Ecological Civilization, develop and improve the theoretical system and discourse system of building an ecological civilization based on China and facing the world and adapt to the new normal, new practice and new requirements in the era of ecological civilization transformation. As such, CASS Thinktank for Eco-civilization Studies, on the basis of carrying out the study of Xi Jinping Thought on Ecological Civilization for a New Era, publishes *Theoretical Construction and Practical Research on Building Ecological Civilization Learning Xi Jinping Thought on Socialism with Chinese Characteristics for a New Era* (hereinafter referred to as *Theoretical Construction and Practical Research on Building Ecological Civilization*), and it is one of the series of "Studies on Xi Jinping Thought on Socialism with Chinese Characteristics for a New Era," a project of the National Social Science Fund of China for the study of new philosophies, new ideas and new strategies of the CPC Central Committee's governance since the 18th CPC National Congress. This is a milestone in the theoretical research of building an ecological civilization and our historic responsibility.

The book focuses on the theme of building an ecological civilization of socialism with Chinese characteristics for a new era and thoroughly studies and explains a series of new philosophies, ideas and strategies related to building an ecological civilization and a community with a shared future for humanity put forward by President Xi Jinping and the Party Central Committee with Comrade Xi Jinping at its core since the 18th CPC National Congress, especially those in the report of 19th CPC National

Congress. It thoroughly and systematically explains, summarizes and refines the theoretical basis, theoretical system, logical starting point and methodology of these new philosophies, ideas and strategies, refines new theories with academic rationality and summarizes new practices with regularity. The book provides a theoretical system, value compliance and practical guidance for promoting the harmonious coexistence between man and nature to proceed from China's realities, develop a global vision and implement 2030 Agenda for Sustainable Development and the Paris Agreement. The authors take the 18th CPC National Congress as an important moment. First, the authors strive to be faithful to the original words, original appearance and original text of Xi Jinping's discourse on ecological civilization and Marx and Engels' dialectics of nature. The authors strive to be faithful to the basic consensus and general proposition of the theoretical and practical achievements made by the CPC Central Committee with Comrade Xi Jinping at its core in uniting and leading the people into the building of socialist ecological civilization since the 18th CPC National Congress. Second, the authors strive to reflect the new trend of socialist ecological civilization toward a new era, new paths and new features of economic and social development entering a new normal and the new vitality of Marxism. Finally, the authors strive to build a theoretical system based on Xi Jinping Thought on Ecological Civilization.

There are ten chapters in *Theoretical Construction and Practical Research on Building Ecological Civilization*. Chapter 1 points out that building an ecological civilization is an issue of a new era, a product of the development of industrial civilization to a certain historical stage and a historical trend of human civilization development. Building an ecological civilization is a major event concerning the five-sphere integrated plan, the four-pronged comprehensive strategy and the realization of the Chinese Dream of national rejuvenation. This chapter stresses that in the new historic juncture in China's development—socialism with Chinese characters for a new era, to address the contradiction between unbalanced and inadequate development and the people's ever-growing needs for a better life, we need to respect nature, follow its ways and protect it. In this connection, we can build a shared future for all life featuring harmonious coexistence between man and nature and fulfill new requirements for a new era.

Chapters 2–9 explain significant theoretical research and scientific judgments of Xi Jinping Thought on Ecological Civilization and draw on theories of economics. "Lucid Waters and Lush Mountains are Invaluable Assets" (Chap. 2), "Environment is People's Livelihood, and Environment is Productivity" (Chap. 3), "Ensuring the Needs for Ecology and Development" (Chap. 5), "Adhere to Ecological Priority and Green Development" (Chap. 6) and "Making Scientific Planning and Implementing the Blueprint Thoroughly" (Chap. 8) aim to explain the internal uniformity of building an ecological civilization and economic development. "Providing Legal Guarantee and Institutional Support for Ecological Progress" (Chap. 7) aims to explain how the building of an ecological civilization integrates into political building and promotes the formation of the new model of modernization with humans developing in harmony with nature in the areas of "rule of law in China" and "China's governance." "Harmonious Coexistence between Humans and Nature Must Be Promoted" (Chap. 4) aims to clarify how to identify the strength and soft power of culture from the perspective of the integration of economy, culture and ecological civilization. This chapter requires people to pay more attention to the traditional ecological wisdom of the Chinese civilization and cultivate and develop the ecological civilization thought of core socialist values and build ecological civilization in contemporary China. "Building a Green Home, Persisting in Promoting the Building of a Community with a Shared Future for Mankind" (Chap. 9) aims to illustrate that the Party Central Committee with Comrade Xi Jinping at its core considers the relation between China's development and the world and the role of China's development in the adaptation to China's status as a large developing country. This chapter indicates that China will remain committed to leading global ecological civilization transformation, promote green growth and shoulder the responsibility of building a community with a shared future for humanity.

Conclusion illustrates the theoretical system of Xi Jinping's discourse on ecological civilization which represents a significant theoretical achievement. On the one hand, this part summarizes and stresses basic scientific judgments of socialist ecological progress established by Xi Jinping Thought on Ecological Civilization and extends to the systematic, historical, continuous and modern characters of Xi Jinping Thought on Ecological Civilization. Therefore, we can deeply understand how the CPC Central Committee with Comrade Xi Jinping at its core has made profound ideological exposition on building an ecological civilization in the twenty-first century as a classic Marxist writer.

It should be noted that *Theoretical Construction and Practical Research on Building Ecological Civilization* is based on the new historical juncture of socialist ecological progress when China's ecological civilization has crossed the threshold into a new era, and China has entered a new normal in economic and social development. This book focuses on studying and expounding on the discourses and a series of speeches made by President Xi Jinping on building an ecological civilization since the 18th CPC National Congress. It should be noted that Xi Jinping Thought on Ecological Civilization has been enriched, developed and improved and has experienced a process from natural gestation to maturity in practice. For example, in 2003, Xi Jinping, then Secretary of Zhejiang Provincial Party Committee, discussed the scientific judgment that "the rise or fall of a civilization is closely tied to its relationship with nature" in *The Rise of a Civilization is Closely Tied to Sound Ecology—Promoting Ecological Progress and Building "Green Zhejiang"* in *Qiushi*. In this article, President Xi Jinping pointed out that ecological environment building is "the objective need of protecting and developing productive forces," "the foundation of socioeconomic development" and "an important symbol of social civilization progress." In 2005, President Xi Jinping put forward the scientific judgment that "Lucid waters and lush mountains are invaluable assets" (which is discussed in *Viewing Ecological Environment from the Perspective of "Two-mountain Theory"*). The logic structure and emotional bond of *Theoretical Construction and Practical Research on Building Ecological Civilization* are based on the thorough analysis and study of the history of Xi Jinping Thought on Ecological Civilization.

President Xi Jinping points out: "This is an era that needs theory and thought and can certainly produce theory and thought." We need to study and implement Xi Jinping's discourses on building an ecological civilization and the spirit of the report of the 19th CPC National Congress and to study and research the historical background, ideological implications, spirit essence and practice requirements of Xi Jinping's discourses on building an ecological civilization comprehensively, accurately, systematically and scientifically. We will take Xi Jinping's discourses on building an ecological civilization as a fundamental instruction, spirit support and practice guideline. We will promote the society to recognize the building of an ecological civilization from aspects of theory, system, path and culture. These are the strategic needs for entering into a new era, building an ecological civilization and realizing the Chinese Dream of national rejuvenation. These are also the responsibilities and missions worthy of the party, the people and ecological civilization theorists should take when they study, research and grasp Xi Jinping Thought of Ecological Civilization.

China's ecological civilization building will lead the world's transformation. China will, building on having basically finished the building of a moderately prosperous society in all respects in 2020 and achieved the first centenary goal, work hard for a further 15 years and basically realize socialist modernization. To the middle of this century, China will be developed into a great modern socialist country that is prosperous, strong, democratic, culturally advanced, harmonious and beautiful. China's modernization is one characterized by harmonious coexistence between man and nature. A great modern socialist country features sound ecology, a new model and sustainable development. Xi Jinping Thought on Ecological Civilization embodies the great wisdom of Chinese philosophy and the brilliance of Marxism. The research and theoretical framing of Xi Jinping Thought on Ecological Civilization is an attempt made by the authors' team standing on the shoulders of giants to learn from President Xi Jinping's discourses on ecological progress. Limited by the team's knowledge, this book needs to be improved in ideological level and theoretical horizon. We urge colleagues from all walks of life to favor us your instruction and criticize and correct us.

Beijing, China Jiahua Pan

Contents

Chapter 1
Scientific Understanding and Theoretical System of a New Era of Ecological Civilization

The report of the 19th CPC National Congress pointed out: Since our 18th National Congress, changes both in and outside China, and the progress made in all areas of China's endeavors, have presented us with a profound question—the question of an era. Our answer must be a systematic combination of theory and practice and must address what kind of socialism with Chinese characteristics the new era requires us to uphold and develop, and how we should go about doing it. This involves fundamental issues like the overarching objectives, tasks, plan, and strategy for upholding and developing socialism with Chinese characteristics in the new era; like the direction, model, and driving force of development, and the strategic steps, external conditions, and political guarantees. The report of the 19th CPC National Congress also pointed out: Building an ecological civilization is vital to sustain the Chinese nation's development. We must realize that lucid waters and lush mountains are invaluable assets and act on this understanding, implement our fundamental national policy of conserving resources and protecting the environment, and cherish the environment as we cherish our own lives. We will adopt a holistic approach to conserving our mountains, rivers, forests, farmlands, lakes, and grasslands, implement the strictest possible systems for environmental protection, and develop eco-friendly growth models and ways of life. We must pursue a model of sustainable development featuring increased production, higher living standards, and healthy ecosystems. We must continue the Beautiful China initiative to create good working and living environments for our people and play our part in ensuring global ecological security. "How to understand ecological civilization, why we need to develop ecological civilization and how to develop ecological civilization" are issues of the times that need to be answered systematically by the theoretical system of building an ecological civilization. They are also fundamental and systematic issues related to the basic understanding of the development stage, basic position, strategic measures, construction mission and development goals of building an ecological civilization. Since the 18th CPC National Congress, General Secretary Xi Jinping has taken the new requirements of the era and practice into consideration, borne people's new aspirations in mind, and made

J. Pan, *China's Global Vision for Ecological Civilization*,
https://doi.org/10.1007/978-981-16-4534-1_1

a series of discourses on building an ecological civilization. Therefore, the systematic and complete Xi Jinping Thought on Ecological Civilization for a New Era has been formed. The Thought systematically describes the theory and practice of socialist ecological civilization contribution in the new era. It represents a new political declaration and practice guideline of consolidating and deepening the new model of modernization with humans developing in harmony with nature.

1 Ecological Civilization is the Historical Trend of the Development of Human Civilization

The opposites of "civilization" are "ignorance" and "barbarism". "Civilization" refers to a progressive state in the development of human society and a product of the advanced stage of human society. The main symbol is, firstly, the invention of writing. American scholar Morgan said: "Without literary records neither history nor civilization can properly be said to exist" [2]. Secondly, the smelting of iron and usage of ironware. Engels pointed out: "The period when all civilized peoples have their Heroic Age: the age of the iron sword, but also of the iron plowshare and ax. Iron was now at the service of man" [3]. He also said that human society begins with the smelting of iron ore and passes into civilization with the invention of alphabetic writing and its use for literary records.

Based on the stage characteristics of production mode, human civilization has gone through three stages: primitive civilization, agricultural civilization and industrial civilization.

In primitive civilization, human beings relied on the nature or used natural objects as their means of livelihood. Hunting and gathering were the most important productive labor, and fire, stone tools and bows and arrows were vital tools for earning a living. The method of making fire by piling and drilling wood invented by Suiren was popularized, and human beings learnt how to cook food. In the primitive times when people were ignorant, there was no decisive boundary between human beings and nature, and human beings adapted to nature like animals and lived like animals. Marx once explained this. He said: "The sole equality to be found in the actual life of animals is the equality between one animal and other animals of the same species; it is the equality of the given species with itself, but not the quality of the genus. The animal genus itself is seen only in the hostile behavior of the different animal species, which assert their particular distinctive characteristics one against another. In the stomach of the beast of prey, nature has provided the battlefield of union, the crucible of closest fusion, the organ connection the various animal species" [4]. After a long and arduous struggle with nature, human beings triumphed over nature, separated themselves from animals and nature, and gradually produced self-centered consciousness. On the whole, people used to passively adapt to nature, work from dawn to dusk, rely on the nature to live, and blindly worship and obey nature. At that time, human beings made no damage and pose no threat to the natural ecology, and were bound by nature in many respects, forming a simple consciousness of primitive civilization.

In agricultural civilization, the main production activities of human beings were farming and animal husbandry. The use of bronze ware, pottery and iron tools, especially the use of iron tools, made human production activities actively turn to the field of productivity development, and human beings began to explore ways and methods to obtain the maximum labor results. Besides, for their survival and development, human beings actively launched a challenge to the earth and began to conquer and transform nature consciously and unconsciously. However, nature also began to retaliate against human beings. Disasters such as drought, waterlogging, mountain torrents and sandstorms continually hit human beings, but they failed to fundamentally threaten people's survival and development.

In the era of industrial civilization, in the 1780s, the British industrial revolution, marked by the use of Jenny spinning machine and Watt steam engine, created the mode of mass production with machines. Marx pointed out: "The steam mill (gives you) society with industrial capitalist" [5]. In the history of mankind, friction fire turns mechanical motion into heat, and the invention and use of steam engine turn heat into mechanical motion. This is a great change in human's understanding and utilization of natural forces, and a sign that material production has entered the mechanized era. In this connection, Engels said that the steam engine is the first real international invention. With the development of machine industry, the bourgeoisie, based on machine production, has expanded their business into mining, energy and raw material production, petrochemical industry, metallurgy and metal processing industry, automobile and aircraft manufacturing and transportation, construction, healthcare and service industries. As such, human beings enter the era of mechanization, automation, electrification and modernization, lead a modern social life with great material wealth, and form an industrial civilization based on the philosophy of "man as master of nature". The civilization emphasizes the conquest of nature by human beings, and sets laws for the earth and world with the attitude of anthropocentrism. Therefore, "there have started into life industrial and scientific forces, which no epoch of the former human history had ever suspected" [6].

However, as Marx pointed out: "On the one hand, capitalist production dramatically develops social productivity. On the other hand, capitalist production is incompatible with the social productivity it produces. The history of capitalist production is the history of confrontation, crisis, conflict and disaster" [7]. During the twenty or thirty years around the middle of the twentieth century, environmental pollution incidents that shocked the world occurred frequently all over the world. Among them, there are eight most serious environmental disasters (including 1930 Meuse Valley fog in Belgium, 1948 Donora smog in America, Great smog of London in Britain, Los Angeles photochemical smog, the outbreak of Minamata disease in Japan, the outbreak of itai-itai disease in Japan, the outbreak of asthma in Yokkaichi, Japan, and the outbreak of Yusho disease in Japan). For example, in Los Angeles, USA, from May to October, 1943, photochemical smog produced by a large amount of automobile exhaust gas caused many residents to suffer from diseases such as red eyes, laryngitis and worsening respiratory diseases, and more than 400 senior citizens over 65 years old died. In the 1948 Donora smog in October 1948, sulfur dioxide and other oxides in the air of Donora mixed with fog, producing sulfuric acid smoke and causing serious air pollution. The smog made 42% of residents fall ill and killed 17 people within 4 days. The Great Smog of London in December 1952 caused more than 4000 deaths. From 1955 to 1968, people living in the Toyama Plain, Japan, suffered from Itai-itai disease because they had river water and rice containing cadmium, and 207 people died.

In the past 300 years, under the guidance of the core value that man can conquer nature, industrial civilization has made fundamental changes in production tools and production methods, thus changing the most revolutionary one among the various components of productivity. In the industrial civilization society, socialized production is like a huge machine. It produces formidable energy day and night, creates hundreds of millions of times of material wealth since the existence of mankind, but also creates mounting and irresistible tension between man and nature. In the past half century, western developed countries have carried out strict environmental protection legislation and law enforcement, and invested much capital and science and technology in building a large-scale environmental protection industry to purify waste. Besides, by upgrading high-end industries and transferring low-end industries to developing countries, these countries achieved so-called "local improvement" or self-proclaimed superiority of sound environment. However, when environmental problems are manifested in the form of ecological system, local actions and improvements are ineffective. At present, human beings are approaching the tipping point of environmental degradation. Regional environmental pollution and large-scale ecological destruction have occurred, and large-scale and global ecological crises such as ozone layer destruction, acid rain, species extinction, land desertification, forest decline, transboundary pollution, marine pollution, sharp decline of wild species, tropical rain forest reduction and soil erosion have seriously threatened the survival and development of all mankind. From a technical perspective, the reason lies in the fact that the production process which should be unified is divided into two relatively independent parts: one part of equipment is used for product production, and the other part of equipment is used for waste purification. Practice shows that the production, construction and operation of purification facilities require huge investment, incur problems of treating diffused pollution, and cause secondary consumption of resources and energy and secondary environmental pollution. From an ethical perspective, human beings share the earth. To pursue the maximum profit of capital, it is neither fair nor moral to protect the environment only in some countries or regions. To solve this basic contradiction and overcome the overall crisis and achieve sustainable development, we must promote the building of ecological civilization. "It's not feasible for us to follow the old path of the US and Europe" [8]. To build a modern country, we need to build an ecological civilization which transcends industrial civilization. This is a comprehensive social transformation and a glorious historical mission.

The rise of ecological production is the inevitable result of the development of modern science and technology revolution. Science and technology have become the first element of modern productivity development and economic growth. The productivity development and economic growth of traditional industrial economy mainly depend on the input of labor, capital and natural resources. In modern society, ecological technology, recycling technology, system management science, complex system engineering, and clean energy and environmental protection industry technology are increasingly becoming the decisive factors of productivity development and economic growth. Qian Xuesen, an outstanding scientist of China, defined the new technological revolution with information industry technology as the core as the fifth industrial revolution from the perspective of the great span of the development of the whole human society. He also put forward that "the core of the fifth industrial revolution is the information problem", "we should not only embrace the fifth industrial revolution, but also make efforts for the fourth industrial revolution" and "foresee the sixth industrial revolution" [9].

In terms of the historical evolution of human civilization, every great development and change of production mode follows the replacement rule, historical blending and natural transformation of civilization and ecology. General Secretary Xi Jinping pointed out: "Historically, the rise or fall of a civilization is closely tied to its relationship with nature" [10]. There are many examples in ancient and modern China and abroad. Engels profoundly pointed out in *Dialectics of Nature*: "We should not indulge in our victory over nature. For every such victory, nature retaliates against us" [11]. "Residents of Mesopotamia, Greece, Asia Minor and other places destroyed forests to obtain land for farming, but it never occurred to them that these places would become barren today" [12]. Mesopotamian civilization (also known as Two Rivers civilization), the earliest civilization in West Asia, was gradually buried by dust and finally forgotten by people. The lessons of history are worth pondering. Nowadays, environmental pollution, ecological destruction and resource shortage are also global public hazards that threaten human survival. We have a long way to go in ecological environment protection. If we don't pay close attention to this work from now on, the price will be even higher in the future. The Chinese nation has a continuous civilization history of more than 5000 years and has made indelible contributions to the progress of human civilization. We must follow the historical laws of ecology and civilization development, vigorously promote the building of an ecological civilization, and be responsible for Chinese civilization and human civilization.

2 Socialist Ecological Civilization is a New Form of Human Civilization Development

Socialist society is a society with development in all respects. The 19th CPC National Congress announced that socialism with Chinese characteristics has entered a new era. The Chinese nation, which since modern times began had endured so much for so long, has achieved a tremendous transformation: it has stood up, grown rich, and is becoming strong. The Communist Party of China, the core of the leadership of the Party and the country, has made more efforts than ever before to "meet the ever-growing economic, political, cultural, social, and ecological needs of our people, and to promote well-rounded human development and all-round social progress". By doing so, "China's will new heights in every dimension of material, political, cultural and ethical, social, and ecological advancement" [14]. According to Marxism, human society is a social organism composed of various parts which are interrelated and interacted with each other, and constantly develops from a lower level to a higher level, which is a natural historical process dominated by certain laws. Socialist society is such a social organism. In our society, people and nature, society and ecology coexist harmoniously. It is a social development form in which social system, natural system and different fields contained in social system adapt to and promote each other and develop together. It is also a social development form formed by the mutual cooperation, promotion and influence of economy, politics, culture and ecology. In our society, economic growth provides material foundation, political progress provides political guarantee, cultural development provides spiritual power and intellectual support, social progress provides sound social environment and conditions, and ecological conservation provides survival foundation and conditions.

Since the 18th CPC National Congress, General Secretary Xi Jinping has put forward a brand-new scientific judgment on the relationship between ecological civilization and socialism. That is, the building of an ecological civilization is an important part of realizing the overall layout of the five-sphere integrated plan for building socialism with Chinese characteristics, an important part of the four-pronged comprehensive strategy which serves as the general strategy of the CPC Central Committee for governing the country and an important part of realizing the Chinese Dream of national rejuvenation. We will develop China into a great modern socialist country that is prosperous, strong, democratic, culturally advanced, harmonious, and beautiful. The five-sphere integrated plan highlights the strategic position of building an ecological civilization and how to understand the socialist ecological progress. The four-pronged comprehensive strategy highlights the strategic measures of building an ecological civilization and how to build an ecological civilization. The great vision of the Chinese Dream highlights the strategic mission of building an ecological civilization and the reasons why we should build ecological civilization. Developing China into a great modern socialist country that is prosperous, strong, democratic, culturally advanced, harmonious, and beautiful not only shows the due position of building an ecological civilization in the overall goal of socialist modernization, but also highlights the great goal and realization vision of building an ecological civilization.

2.1 Important Content of the Five-Sphere Integrated Plan for Building Socialism with Chinese Characteristics

Since the 18th CPC National Congress, the Party Central Committee with Comrade Xi Jinping at its core, focuses on the general basis of the primary stage of socialism and the organic unity of the realization of socialist modernization and the overarching goal of the great rejuvenation of the Chinese nation. The Party Central Committee with Comrade Xi Jinping at its core has repeatedly emphasized that we must adhere to the five-sphere integrated plan for building socialism with Chinese characteristics, including the building of an ecological civilization. It has demanded that the deterioration of the ecological environment be reversed from the source, create a sound production and living environment for the people, and strive to build Beautiful China and realize the sustainable development of the Chinese nation. Regarding the strategic significance of integrating ecological progress into the five-sphere integrated plan for building socialism with Chinese characteristics, General Secretary Xi Jinping pointed out: "The 18th CPC National Congress integrated ecological progress into the five-sphere integrated plan for building socialism with Chinese characteristics, and clearly proposed that we must vigorously promote ecological progress and strive to continue the Beautiful China initiative to realize the sustainable development of the Chinese nation. This marks the further deepening of our understanding of the laws of socialism with Chinese characteristics and shows our firm will and strong determination to strengthen the building of an ecological civilization" [14].

Socialism and ecological civilization are highly consistent. The proposal of ecological civilization is based on Marxism's complete and scientific grasp of the overall historical process of human society, and is intrinsically and logically unified in the essence of socialism. Socialist ecological civilization originates from the inherent consistency in socialist economic, political and ecological progress, and from the fact that socialism can maximally follow the harmonious development law between man, nature and society. As Marx said: "This kind of communism, as completed naturalism, is equal to humanitarianism. The completed humanitarianism is naturalism and the real solution to the contradiction between man and nature, and between man and man and to the struggle between existence and essence, objectification and self-confirmation, nature and necessity, and between individuals and classes". Socialist ecological civilization represents a new form of human civilization development. The essence of socialism empowers socialism to surpass capitalism. In a socialist society, the Party and government that represent the people in power are not the representatives of any interest group, but the fundamental interests of all the people. Socialism transcends concrete, immediate and local interests, takes a long-term approach to human civilization development, and will unite, guide and lead all the people to embrace the bright future of human society.

Practice is a unique objective activity of human beings and the way of human survival. Marx believes that, because of the integration of practice factors, nature has become social and historical nature, that is, humanized nature. At present, China is facing two major contradictions in its development: one is the contradiction between underdeveloped economy and people's growing material and cultural needs, which will be a long-term major contradiction, and the solution to this contradiction depends on development. The other is the contradiction between economic and social development and increasing pressure on population, resources and environment, which is becoming increasingly prominent now. It is a strategic consideration to solve this contradiction to integrate the building of an ecological civilization into the five-sphere integrated plan for building socialism with Chinese characteristics. Since the reform and opening up, China has remained committed to economic development to promote rapid economic growth. In this process, we emphasize sustainable development, energy conservation, emission reduction and environmental protection. However, in some places and fields, the relationship between economic development and ecological environment protection has not been properly handled. Economic development has been achieved at the expense of uncontrolled consumption of resources and environmental damage, resulting in the shortage of energy resources and increasingly prominent ecological and environmental problems. For example, the constraints on energy resources have been strengthened, and the foreign dependence of important resources such as oil has risen rapidly. The area of farmland approaches the red line of 1.8 billion mu, and soil erosion, land desertification and grassland degradation are serious. Some areas have approached or exceeded the limit of environmental carrying capacity due to blind development, over-exploitation and disorderly development. Some areas of China are hit by smog frequently, and various environmental pollution such as air pollution, water pollution and soil pollution are on the rise. If this situation does not change, the energy resources will be difficult to support our development and the ecological environment will be overwhelmed, which in turn will inevitably exert a serious impact on the sustainable development of economy and the space and momentum for China's development.

From a global perspective, the rapid development of western industrialization creates huge material wealth, but western countries pay a very heavy ecological and environmental cost. The lessons the incidents left us with should never be forgotten. Incidents happened in the twentieth century, such as the photochemical smog incident in Los Angeles, the Great Smog of London, and the Minamata disease incident in Japan, had a great impact on the ecological environment and people's life. In some countries and regions, such as heavy metal polluted areas, water and soil have been polluted, and the environment is difficult to restore. General Secretary Xi Jinping pointed out: "Many countries in the world, including some developed countries, have followed the beaten path of 'pollution first, treatment afterwards'. They destroyed the ecological environment when pursuing development, and created a lot of worthless or even destructive things. To make up for it, the cost is more than the wealth created at the beginning" [17]. China is a major developing country. It is not feasible to follow the beaten path of "pollution first, treatment afterwards" to build a modern country. We must identify a new way to protect environment.

Ecological, economic, political, cultural and social progress constitutes the five-sphere integrated plan. This is a qualitative leap from the traditional pattern of the overall layout of socialism with Chinese characteristics, which is composed of economic, political, cultural and social development, and makes the strategic position of building an ecological civilization in the overall layout of socialism with Chinese characteristics undergo fundamental and historic changes. It shows that, to build socialism with Chinese characteristics, we should not only promote coordinated progress in the socialist economic, political, cultural, social and eco-environmental fields, but also fully implement the prominent position of building an ecological civilization. We must strengthen coordinated innovation, promote the coordinated development of socialist material, political, cultural, social and ecological civilizations, and enrich the overall plan for building socialism with Chinese characteristics.

2.2 Important Content of the Four-Pronged Comprehensive Strategy for Building Socialism with Chinese Characteristics

General Secretary Xi Jinping pointed out that building an ecological civilization is an important part of the four-pronged comprehensive strategy. The overall plan answers the question of what is socialism, and the overall strategy solves the problem of how to build socialism. The four-pronged comprehensive strategy is not only strategic and development goals, but also strategic initiatives and strategic key points. Logically speaking, building a moderately prosperous society in all respects is the overall goal put forward by the 18th CPC National Congress. Comprehensively deepening reform and advancing law-based governance are two wings, and exercising full and rigorous governance over the Party is the core of leadership. Each one covers the building of an ecological civilization, and reflects the inherent requirements and goals of building an ecological civilization.

First, we will build a moderately prosperous society in all respects and strive to embody that "ecological environment is the key of a moderately prosperous society". The quality of ecological environment is a very important weight to measure the construction of a moderately prosperous society. We shouldn't develop economy at the cost of people's life quality. In recent years, some prominent environmental problems have made negative impacts and harmful consequences to people's production, life and health. We used to emphasize "production development, sound ecology and happy life". Now we must integrate "health" with them. We must pursue a civilization development path featuring a win–win for production, ecology, life and health, which is a sustainable lifestyle and a higher-level social civilization structure that is more consistent with the essential characteristics of a moderately prosperous society.

Second, we will comprehensively deepen reform, and develop institutions and mechanisms for ecological civilization. To build ecological civilization, we must establish a systematic and complete system of ecological civilization institution to protect the ecological environment with a system. General Secretary Xi Jinping emphasized that it is necessary to deepen the reform of ecological civilization system, establish the "multiple pillars (which mean the basic structure)" of ecological civilization system as soon as possible, and institutionalize and strengthen rule of law in building an ecological civilization. Since the 18th CPC National Congress, a series of target systems, assessment methods, reward and punishment mechanisms that cover and reflect the requirements of ecological civilization, and a series of institutional arrangement documents have been introduced, including systems for developing and protecting territorial space, farmland protection systems, systems of water resource management, systems for environmental protection, the compensation system for use of resources, ecological compensation systems and environmental damage compensation systems. These documents include *Ecological Conservation Attainment Evaluation Methods, Guidelines for the Construction of Unified and Standardized National Pilot Ecological Civilization Zones and Opinion on Comprehensively Promoting the River Chief System.* These are all institutional changes which comprehensively deepen reform and lead the building of an ecological civilization. They have far-reaching effects and reflect the profound sense of innovation and historical responsibility of the Party Central Committee with Comrade Xi Jinping at its core.

Third, we will advance law-based governance in all areas, and strengthen rule of law in ecological progress with the strictest system and rule of law. In recent years, a series of laws and regulations and interpretations, such as *Environmental Protection Law of People's Republic of China, Law of the People's Republic of China on the Prevention and Control of Atmospheric Pollution, Amendment to Law of the People's Republic of China on Prevention and Control of Water Pollution (Draft), Environmental Protection Tax Law of the People's Republic of China,* and *Interpretation of the Supreme People's Court and the Supreme People's Procuratorate on Several Issues concerning the Application of Law in the Handling of Criminal Cases of Environmental Pollution* have been introduced, revised and interpreted. This has embodied the legal consciousness of the Party Central Committee with Comrade Xi Jinping at its core. They are guided by law-based governance in all areas and make efforts to ensure sound lawmaking, strict law enforcement, impartial administration of justice, and the observance of law by everyone in the field of building an ecological civilization.

Fourth, we will exercise full and rigorous governance over the Party, completely transform the view of achievement, and provide the people with the fairest ecological public products and the most inclusive people's livelihood and well-being. T We must strengthen the Party's leadership over the building of an ecological civilization by overhauling the achievement view of GDP-first and leveraging the system of assessment. For example, *Ecological Civilization Construction Evaluation and Assessment Methods* combines evaluation and assessment. Evaluation focuses on guidance, the weights of indices such as ecological environment quality and public satisfaction that reflect the people's personal feelings are improved in the index system. Assessment focuses on constraints, rewards and punishments, and results are taken as an important basis for the comprehensive evaluation, rewards, punishments, appointments and removal of government and Party officials and leadership teams. This has played a very important guiding role in forming a new view of achievement, guiding and urging government and Party officials at all levels to promote ecological conservation consciously. Under the new view of achievement, government and Party officials take responsibility for building an ecological civilization in performing their duties, and they cooperate with departments.

2.3 Building an Ecological Civilization is an Important Part of Realizing the Chinese Dream of National Rejuvenation

General Secretary Xi Jinping pointed out: "Building an ecological civilization is an important part of realizing the Chinese dream of national rejuvenation" [18]. Realizing the Chinese dream of national rejuvenation is a long-cherished wish and common pursuit shared by the whole Party and the Chinese people. Today, we are closer, more confident, and more capable than ever before of making the goal of national rejuvenation a reality. The 18th National Congress of the Communist Party of China proposed to build a moderately prosperous society in all respects when the Communist Party of China has been founded for 100 years, and achieve decisive results in in key areas, and general frameworks for reform by 2020. The 19th CPC National Congress put forward that the period between the 19th and the 20th National Congress is the period in which the timeframes of the two centenary goals converge. In this period, not only must we finish building a moderately prosperous society in all respects and achieve the first centenary goal; we must also build on this achievement to embark on a new journey toward the second centenary goal of fully building a modern socialist country. This not only means that China boasts more perfect systems and more vibrant society in all aspects, but also means that China is becoming more open to the outside world, has more affinity, and makes greater contributions to human civilization. Socialist ecological civilization is an all-round change in the history of human civilization. It needs an all-round innovation and exploration from value to culture, from economy to politics. Failure of building an ecological civilization will inevitably exert negative effects on the realization of the Chinese dream. The Chinese Dream emphasizes the historical inheritance of China's civilization which was nurtured over more than 5000 years. This concept will eventually promote the return of contemporary Chinese and global ecological progress to Chinese civilization and traditional ecological wisdom and thought. In the past 170 years, China was plunged into the darkness of foreign aggression, and realized independence and liberation after patriots fought against the odds. The Chinese Dream explores the extraordinary history. The Chinese people were ravaged by war, and saw their homeland torn apart, which resulted in the historical break and temporary pains of the traditional Chinese ecological civilization concept. The Chinese Dream requires us to deeply feel these sentiments and understand the setbacks, complexities and difficulties in China's ecological civilization progress in a calm and rational manner. Today, the Chinese Dream takes the building of an ecological civilization as an important part. This indicates that the great rejuvenation of the Chinese nation featuring respect for nature, adaptation to nature and protection of nature as its civilization tradition will win the respect of concepts, emotional closeness and support in actions of other nations.

2.4 Developing China into a Great Modern Socialist Country that is Prosperous, Strong, Democratic, Culturally Advanced, Harmonious, and Beautiful

The report made by General Secretary Xi Jinping at the 19th CPC National Congress put "beautiful" as an important goal of socialist modernization in a new era into the report of the National Congress of the Party for the first time, and strengthened the overall goal of socialist modernization "prosperous, strong, democratic, culturally advanced, harmonious, and beautiful" in many places. In the first part of the report "The Past Five Years: Our Work and Historic Change", it is pointed out that we must strive to build China into a great modern socialist country that is prosperous, strong, democratic, culturally advanced, harmonious, and beautiful. In the third part of the report "The Thought on Socialism with Chinese Characteristics for a New Era and the Basic Policy", it is pointed out: The Thought on Socialism with Chinese Characteristics for a New Era makes the following things clear: It makes clear that the overarching goal of upholding and developing socialism with Chinese characteristics is to realize socialist modernization and national rejuvenation, and, that on the basis of finishing the building of a moderately prosperous society in all respects, a two-step approach should be taken to build China into a great modern socialist country that is prosperous, strong, democratic, culturally advanced, harmonious, and beautiful by the middle of the century. In the fourth part of the report "Securing a Decisive Victory in Building a Moderately Prosperous Society in All Respects and Embarking on a Journey to Fully Build a Modern Socialist China", it is once again pointed out: In the second stage from 2035 to the middle of the twenty-first century, we will, building on having basically achieved modernization, work hard for a further 15 years and develop China into a great modern socialist country that is prosperous, strong, democratic, culturally advanced, harmonious, and beautiful.

In the past 30 years, the goal of socialist development adheres to the internal unity of material, political, cultural and ethical civilization. In 1987, the report of the 13th CPC National Congress put forward that "we must strive to develop China into a modern socialist country that is prosperous, strong, democratic and culturally-advanced". In 207, the report of the 17th CPC National Congress put forward that "we must strive to build China into a modern socialist country that is prosperous, strong, democratic, culturally-advanced and harmonious". Building China into a great modern socialist country that is prosperous, strong, democratic, culturally advanced, harmonious, and beautiful highlights the new expansion of the overall plan for building socialism with Chinese characteristics by ecological civilization, Beautiful China and harmony between man and nature, represents the inevitable requirement of coordinated implementation of the five-sphere integrated plan and the four-pronged comprehensive strategy, and shows the goal and development momentum of building an ecological civilization in the process of realizing the Chinese Dream of national rejuvenation. In other words, realizing the Chinese Dream of national rejuvenation must also be a beautiful Chinese dream of national rejuvenation [19].

3 Theoretical System of Building a Socialist Ecological Civilization

General Secretary Xi Jinping attaches great importance to building an ecological civilization. Since the 18th CPC National Congress, General Secretary Xi Jinping has delivered a series of discourses on ecological civilization and its progress, made a large number of special discussions and important instructions, and put forward many brand-new scientific judgments featuring philosophical thinking, economic rationality, humanistic feelings and global vision. With affectionate and simple words, fresh and natural style of writing, profound and rich implications, Xi Jinping Thought on Ecological Civilization for a New Era answers a series of major theoretical and practical problems facing the building of an ecological civilization in China and today's world, and formed a scientific and complete theoretical system related to the basic connotation of building an ecological civilization, reasons for and ways to build ecological civilization. This provides scientific guidance for moving towards a new era of socialist ecological civilization, realizing the beautiful Chinese dream of national rejuvenation, and promoting the building of an ecological civilization community with a shared future for mankind.

Xi Jinping Thought on Ecological Civilization for a New Era contains a series of new scientific judgments guiding how to enter a new era of socialist ecological civilization. For example, with regard to the basic concept of ecological civilization, General Secretary Xi Jinping has repeatedly stressed the need to "observe the laws of nature, respect nature, follow its way and protect it" and "continuously promote the harmonious development between man and nature". In terms of the development rule of ecological civilization, General Secretary Xi Jinping pointed out: "Historically, the rise or fall of a civilization is closely tied to its relationship with nature" [20]. Looking at ecological civilization from the perspective of political economy, General Secretary Xi Jinping clearly pointed out: "Protecting ecological environment means protecting productive forces, and improving ecological environment means developing productive forces" [21]. "We want to have not only mountains of gold, but also mountains of green. If we must choose between the two, we would rather have the green than the gold. And in any case, green mountains are themselves gold mountains" [22]. In terms of development stages of ecological civilization, General Secretary Xi Jinping pointed out: "Ecological civilization is a significant achievement in the progress of human society. Human beings have experienced primitive civilization, agricultural civilization and industrial civilization. Ecological civilization is the product of the development of industrial civilization to a certain stage and a new requirement for the harmonious development of man and nature" [23]. On the relationship between ecological civilization and socialism, General Secretary Xi Jinping pointed out that ecological progress is an important part of the five-sphere integrated plan and the four-pronged comprehensive strategy. Regarding the path of building an ecological civilization, General Secretary Xi Jinping pointed out: "We must put ecological civilization progress high on the agenda and integrate it into economic, political, cultural, and social advancement" [24]. In terms of institution building in ecological civilization, General Secretary Xi Jinping pointed out: "To protect the ecological environment, we must rely on institutions and the rule of law. Only by implementing the strictest system and rule of law can we provide a reliable guarantee for building an ecological civilization [25]. Regarding ecological civilization and China's fine traditional culture, General Secretary Xi Jinping pointed out: "China's civilization was nurtured over more than 5000 years and accumulated rich ecological wisdom" [26]. Regarding the realistic driving force of building an ecological civilization, General Secretary Xi Jinping pointed out: "We must clearly understand the urgent and onerous nature of protecting the ecological environment and controlling environmental pollution, and be soberly aware of the importance and necessity of strengthening the building of an ecological civilization [27]. In terms of the historic mission of building an ecological civilization, General Secretary Xi Jinping pointed out: "Building an ecological civilization is an important part of realizing the Chinese dream of national rejuvenation" [28]. On the global governance of building an ecological civilization, General Secretary Xi Jinping pointed out: "China must promote and strengthen ecological progress from a global perspective" [29] and "become an important participant, contributor and leader in the global endeavor for ecological civilization" [30]. Engels pointed out: "This is the greatest progressive revolution that mankind had so far experienced. It is an era called for giants and produced giants in power of thought, passion and character, in universality and learning" [31]. The scientific conclusions of Xi Jinping Thought on Ecological Civilization embody wisdom of all times from home and abroad. With an open vision and broad mind, it reflects General Secretary Xi Jinping's sincere love for nature, continuous creation enthusiasm and profound ecological knowledge.

Xi Jinping Thought on Ecological Civilization for a New Era is an important part of Xi Jinping Thought on Socialism with Chinese Characteristics for a New Era, which answers fundamental questions concerning the overall ecological progress systematically, comprehensively, scientifically and completely. It reflects that the Party Central Committee with Comrade Xi Jinping at its core unites and leads the whole Party and the Chinese people of all ethnic groups, focuses on the progress of the building of an ecological civilization, the need and new expectations of the people's ecological and environmental protection, overcomes a series of major challenges, and vigorously promotes the building of an ecological civilization to obtain valuable experience of new major achievements. It reflects that the Party Central Committee with Comrade Xi Jinping at its core takes Marxism-Leninism, Mao Zedong Thought, Deng Xiaoping Theory, the Theory of Three Represents, and the Scientific Outlook on Development as the guidance, combines the basic Marxist theory of man and nature and the basic principles of dialectics of nature with the reality of building an ecological civilization in contemporary China and the characteristics of socialist ecological civilization in a new era, and makes major theoretical achievements.

Xi Jinping Thought on Ecological Civilization is the latest development of Marxist view of human and nature and ecological view in contemporary China. It is also a brand-new ecological civilization view, theoretical realm and discourse system of Marxism. The new discourse system has realized the true unity of Marx and Engels' theory on the relationship among human, nature, society, human and nature, human and society, nature and society, and free nature and humanized society. The relationship between human and nature is the primary problem to be solved in Marxist dialectics of nature. Whether human and nature are in conflict or harmony, contradictory or symbiotic is an important question that human beings need to answer, and one of the important criteria to distinguish Marxism from non-Marxism. Marxist dialectics of nature reflects Marxist philosophy of nature and natural science, which embodies the unity of world view, epistemology and methodology of Marxist philosophy and is an important part of Marxist philosophy. Dialectics of nature, as one of the important components of natural philosophy, cannot develop without the development of natural practice. It can be said that the development of dialectics of nature is closely related to the development of natural science. The rapid development of natural science in the twentieth century greatly expanded human understanding of nature, far exceeding the horizon of natural science in the nineteenth century. The development of natural science in the twentieth century has revealed the dialectics of nature and natural science in a wider scope and to a deeper degree. Therefore, many basic viewpoints of dialectics with solid natural science facts are widely accepted by the natural science community. In essence, Xi Jinping Though on Ecological Civilization, like dialectics of nature, is a theoretical system. In particular, the concepts of the unity of development and protection, natural value and natural capital in the concept that clear waters and green mountains are mountains of gold and silver, the concept of space balance, and the concept that mountains, rivers, forests, farmlands, lakes, and grasslands are a community of life, reflect General Secretary Xi Jinping's familiarity with dialectics of nature. Furthermore, General Secretary Xi Jinping's new scientific judgment and theoretical system on building an ecological civilization, just like General Secretary Xi Jinping's discourses on economic, political, cultural, social progress and Party building and related scientific judgments, are an important part of Xi Jinping Thought on Socialism with Chinese Characteristics for a New Era, and become an independent chapter of building an ecological civilization with the brilliance of Marxist view of human and nature. They provide scientific guidance and fundamental compliance for the building of an ecological civilization in the new era of socialism. They also provide Chinese plan and wisdom for guiding the implementation of Transforming our World: *2030 Agenda for Sustainable Development*.

3.1 Connotation System of "What is Ecological Civilization"

The basic connotation of ecological civilization is respecting nature, following its ways, and protecting it. General Secretary Xi Jinping has repeatedly stressed that it is necessary to "firmly establish the ecological civilization concept of respecting nature, following its ways, and protecting it" [29]. He has made clear the basic connotation of ecological civilization from epistemology, methodology and practice theory, fundamentally clarified concepts of ecological civilization and avoided misunderstanding of them. Respecting nature means to make it clear that nature breeds and nurtures human beings, and that nature is the parents of human beings. Just as Engels pointed out: "We, with flesh, blood, and brain, belong to nature, and exist in its midst" [30]. We must follow ways of nature. As Marx said: "Human plans that are not based on laws of nature will only bring disasters" [31]. Protecting nature is to protect the natural environment and natural resources. Respecting nature is related to people's awareness, and following ways of nature is related to methodology. Today, a series of unsustainable economic, political and social development problems caused by environmental destruction, ecological degradation and tightening resource constraints are the consequences caused by human beings' wanton destruction to nature. General Secretary Xi Jinping pointed out: "On the issue of ecological environmental protection, we shouldn't transgress the bounds, otherwise, we will be punished" [32].

Human, mountains, rivers, forests, farmlands, lakes, and grasslands are a community of life. This is the new realm of the development of Marxist philosophy of nature and the theoretical basis of ecological civilization. Marx and Engels founded the Marxist philosophy of nature of dialectical materialism, but this creation process is arduous. When Marxist philosophy was founded in the 1840s, Marx and Engels put forward some basic ideas of dialectical materialist view of nature. In the middle and late periods of 1970s, Engels systematically expounded Marxist philosophy of nature in *The Part Played by Labour in the Transition from Ape to Man*. Engels himself explained the philosophy, and "went through as complete as possible a 'moulting' in mathematics and the natural sciences" [15]. At the beginning of the twenty-first century, during the historical journey of building an ecological civilization and realizing the Chinese dream of national rejuvenation, General Secretary Xi Jinping formed a brand-new scientific conclusion that people are integrated into nature and nature takes precedence over human beings. That is, human, mountains, rivers, forests, farmlands, lakes, and grasslands are a community of life. In terms of the relationship between human and nature, Marx pointed out that "human stands on a stable earth and breathes all natural forces" [16]. General Secretary Xi Jinping's scientific conclusion is a new summary and development of Marx's idea. It urges mankind to re-examine the "anthropocentrism" which has been popular for hundreds of years in industrial civilization. It is the latest realm of Marxist philosophy of nature, embodies great theoretical charm, and becomes the theoretical and cultural basis of ecological progress in contemporary China.

The rise or fall of a civilization is closely tied to its relationship with nature. This is the operation law of ecological civilization from the view of civilization history. The view of civilization history focuses on the historical process of human civilization. According to the view of civilization history, history of the development of human society is essentially history of the evolution of human civilization and interactive history of the relationship between human and nature. The relationship between human and nature interweaves, permeates and transforms, and jointly promotes human civilization from low level to high level. In the history of the development of the four ancient civilizations, ancient Egyptian civilization originated in the lower reaches of Nile River, ancient Indian civilization mainly originated in the Ganges basin, and ancient Babylonian civilization mainly originated in Mesopotamia within the Tigris–Euphrates river system. They used to be places boasting dense forests and abundant water and plants. However, due to deforestation and encroachment on wetlands, civilization fell or migrated after fertile lands became desert. China boasts a civilization which lasted for 5000 years, and agriculture with a long history. They have made a profound impact on the survival and development of the Chinese nation and the creation of civilization. China is one of the earliest origins of agriculture in the world. Millet dryland crops originated in the Yellow River basin, and artificially cultivated rice originated in the middle and lower reaches of the Yangtze River. Today, China is in an important stage of post-industrialization development. Take water pollution and treatment as an example. The Yangtze River, the Yellow River and the Pearl River, which occupy one third of the total area of China, are subjected to serious water pollution. The Communist Party of China leads the people to explore and enter a new era of socialist ecological civilization. This is a historical opportunity and an irreversible historical trend of the development of the times. General Secretary Xi Jinping stressed that "the rise or fall of a civilization is closely tied to its relationship with nature". This revealed the law of human civilization development. In front of the tide of history, we will take it as the fundamental starting point, follow the historical law of ecology and civilization development, protect the ecology. Therefore, we can build ecological civilization in a better way.

3.2 Driving Force System of "Reasons for Building an Ecological Civilization"

Xi Jinping pointed out that we must clearly understand the urgent and onerous nature of protecting the ecological environment and controlling environmental pollution, and be soberly aware of the importance and necessity of strengthening ecological conservation. Problems in ecological civilization drive us to form the national conditions view of building an ecological civilization. Marx said: "Problems are the slogan of the times, and they are the most practical voice to the state" [35]. General Secretary Xi Jinping pointed out: "As the CPC has fought its way through revolution, construction and reform, its sole aim has always been to solve the problems of China" [36].

China now faces arduous tasks and enters a critical stage of development, a challenging stage of reform, and a period in which problems are becoming increasingly prominent. Many problems are intertwined. Comrade Deng Xiaoping once predicted that "there will be more problems after economy prospers than when it is not developed" [37]. In recent years, serious pollution problems, such as smog, unsafe drinking water in some areas and heavy metal contamination of soil, have occurred frequently, and are widely discussed by the whole society. General Secretary Xi Jinping pointed out that it is necessary to clearly understand the urgency and arduousness of protecting ecological environment and controlling environmental pollution, and the importance and necessity of strengthening the building of an ecological civilization. The keys lie in the problems of resources and extensive development. General Secretary Xi Jinping said: "If we still follow the path of extensive development, even if we achieve the goal of doubling the GDP, how will our development affect the environment? I'm afraid it will do damage to the resources and environment". Development will not be achieved if we continue to follow the old path, consume resources and pollute the environment at all cost.

The people's aspiration to live a better life is our goal. People's livelihood highlights the mass view of ecological civilization. To test the effectiveness of all our work, we must see whether the people have really benefited and whether people's lives have been improved. These are the essential requirements of building the Party for the public good and exercising state power for the people, and an important guarantee for advancing the cause of the Party and the people. A sound ecological environment is the fairest public product and the most inclusive welfare of people's livelihood. As people pay more attention to environmental problems, it can be said that the ecological environment will play an increasingly prominent role in the happiness index of people's lives. General Secretary Xi Jinping pointed out: "We must put ecological progress high on the agenda. This is what the people want. People are not dissatisfied with the growth rate of GDP, but with the poor ecological environment. We must figure out what the people want. We must bear people's aspirations in mind. The ecological environment is very significant". "We shouldn't take strengthening ecological conservation and ecological environment protection and promoting green and low-carbon lifestyle as economic issues only. They are all related to politics" [39].

The protection and improvement of ecological environment underlies the protection and development of productive forces. Productive forces highlight the dialectical view of building an ecological civilization. We used to see economic development and ecological environment protection as conflicting goals, so we made some mistakes in practice. Productivity determines the nature of production relations, and shows the relationship between human and nature. Our country is still in the primary stage of socialisms. As we have introduced the reform and opening up for 40 years, the principal contradiction facing Chinese society has evolved. What we now face is the contradiction between unbalanced and inadequate development and the people's ever-growing needs for a better life. Marx believed that human survival and development depend on the material resources of nature. Necessary living materials such as food, clothing, housing and transportation are obtained from nature. Land is a special natural resource and the most basic material for human production and life. However, due to the rapid destruction of the environment and the accelerated degradation of ecology, polluted water flowing on the earth, grain produced on the land and air as a material all pose a threat to the survival and development of mankind. Therefore, productivity that ignores ecological environment protection is not advanced productivity. We must implement General Secretary Xi Jinping's new idea that protecting ecological environment means protecting and developing productive forces, and make ecological environment play a unique role in green, low-carbon and circular development as a factor of productive forces. In this connection, people can realize that growth is a quantitative concept, and sustainable development is a quality concept and real productive force. Only when we attach great importance to the ecological environment as an essential factor of productivity, observe the laws of nature, and protect and utilize the ecological environment, can we better develop productivity and realize the harmonious coexistence between human beings and nature at a higher level.

Lucid waters and lush mountains are invaluable assets. Green industry drives us to highlight the values of building an ecological civilization. General Secretary Xi Jinping pointed out: "We want to have not only mountains of gold, but also mountains of green. We prefer clear waters and green mountains to mountains of gold and silver, and clear waters and green mountains are mountains of gold and silver" [40]. Since the introduction of the reform and opening up over 30 years ago, we have remained committed to economic development to achieve rapid economic growth. However, China used to pursue rapid economic development at the cost of uncontrolled resources consumption and environmental destruction. Therefore, the problems of energy, resources and ecological environment become more and more acute. Due to blind development, over-exploitation and disorderly development, some areas have approached or exceeded the limit of the resources and environmental carrying capacity. River pollution, drinking water safety, soil pollution, frequent large-scale and long-term smog have seriously affected the people's sense of happiness and fulfillment, and restricted the sustainable development of economy and society. A very important reason is that the relationship between economic development and environmental protection has not been properly handled. We failed to understand the strategic position of building an ecological civilization, and neglected the long-term consequences of social production. Engels pointed out: "Up to now, all modes of production only aim at obtaining the latest and most direct benefits of labor. Those distant results that appear later and entail gradual repetition and accumulation are completely ignored" [41]. Based on this, we must adapt to the new normal condition based on the characteristics of China's economic growth in the current phase and stay cool-minded. We must look at ecological civilization from the new perspective of the new normal. Green economy development means that we will not pursue rapid economic growth at the sacrifice of environment protection, and will give up the traditional unsustainable extensive growth model. We need to look at ecological civilization from the perspective of the new normal. Green economy development means that the economic growth rate is appropriate, the structure is optimized, and human and nature develop harmoniously. It can be predicted that green industry and green economy will become the new normal of China's national economic development and become an important opportunity to promote China's transformation from an economic giant to an economic powerhouse.

We will work to build a community with a shared future for mankind. All humans share an ecological home, and that is why we highlight the global view of building an ecological civilization. General Secretary Xi Jinping pointed out: "Mankind, by living in the same global village within the same time and space where history and reality meet, has increasingly emerged as an intimate community of shared future" [42]. Mankind has only one Earth. From a global perspective, since the publication of *Our Common Future* in 1987, sustainable development has been widely recognized, and has become a development strategy shared by all countries in the world. In 2013, China's ecological civilization concept was formally adopted into Decision at the 27th Governing Council of the United Nations Environment Programme. This means that when we face global challenges, we must cooperate to cope with them, and jointly turn pressure into strength and adversity into opportunities. From the perspective of civilization exchange, as early as over 2000 years ago, the Chinese people came to recognize that "it is only natural for things to be different". Greater exchanges and mutual learning among civilizations can further enrich the colors of various civilizations and the cultural life of people, and open up still greater alternatives in the future. We should understand the Chinese history and culture, and open our eyes to the world, understand the history and culture of different nationalities in the world. We need to take the essence and discard the dross to learn from them. We must adhere to the diversity of civilizations, and lead the world trend of mutual tolerance, mutual learning and exchange of civilizations. By doing so, we will do our part for the progress of human civilization, and become an important participant, contributor and torchbearer in in the global endeavor for ecological civilization.

3.3 How to Build a Strategic System of Ecological Civilization

General Secretary Xi Jinping pointed out: "It is necessary to understand the significance of incorporating ecological progress into the integrated plan of China's socialist cause" [44]. We must attach greater importance to ecological civilization. We need to drive economic, political, cultural, and social advancement with ecological progress, and improve the strategic significance of integrating ecological conservation into economic, political, cultural, and social progress.

We must integrate ecological civilization progress into economic development. First, we should act on the principles of prioritizing resource conservation and environmental protection and letting nature restore itself. This is because we face tightening resource constraints, serious environmental pollution and severe ecosystem degradation. Putting conservation first in resources, protection first in environment and natural restoration first in ecology form a unified organic whole, which is the basic strategy and fundamental philosophy of integrating ecological civilization into economic development. Second, we need to more consciously promote green, circular and low-carbon development. This is determined by the basic connotation of building an ecological civilization. Green development focuses on the upgrading of traditional industries and their economies, with more emphasis on changing the development mode and adjusting the industrial structure. The core of circular development is to improve the efficiency of resource utilization. The basic idea is that waste can be turned into resources. The essence of circular development is resource recycling and reuse, and minimize the damage to the environment caused by waste. Low-carbon development is characterized by low-carbon emissions. We need to improve energy efficiency by saving energy, develop renewable energy and clean energy, and increase forest carbon sinks. We need to reduce energy consumption intensity, carbon intensity and total carbon emissions. It is a concept and development state closely related to climate change. This is the basic path for ecological civilization to integrate into economic development, and the only way for the formation of the new normal of China's economy put forward by General Secretary Xi Jinping.

We will remain committed to integrating ecological progress into political advance. First, we will no longer regard GDP growth as the most important criterion. This is the most crucial way to integrate ecological civilization into political building. Officials of Party organizations at all levels should set a good example. They are the backbone of the Party's cause. If we do not change the assessment standard of officials, we can't deter those who do not bear ecological environment in mind and make decisions blindly leading to serious consequences. Some officials destroy the ecological environment, "take up another post and do not hold accountable for what they have done" [44]. This is dangerous and practice has proved it. Second, we must strengthen top-level design and overall planning. This is the key to the integration of ecological civilization into political progress. Strengthening the top-level design of ecological civilization means that we need to carry out overall design, coordination and promotion of the reform of ecological civilization system. This is the political guarantee for the Party to lead the building of ecological civilization. Third, we will promote the institutional and legislative reform of ecological civilization system and the regularization of the innovation and rule of law culture of ecological civilization system. This is the foundation of integrating ecological civilization into political progress.

We will insist on integrating ecological progress into cultural advancement. First, we will constantly form a new lifestyle. This is a turning point in the concept of integrating ecological civilization into cultural advancement. In China, excessive consumption, conspicuous consumption and wasteful consumption are affecting people's inner happiness, and make people fall into materialism. The implementation of the eight-point frugality code constrains officials from pursuing luxurious consumption. With a huge population and limited farmland resources, it is necessary to form a simple, low-carbon, fair and green lifestyle. Second, we will improve the soft power of the Chinese culture. This is the cultural guarantee of the integration of ecological civilization into cultural advancement. Confucianism, Buddhism and Taoism have represented the mainstream of the traditional Chinese culture for 5000 years. They put forward that people should respect nature and embrace other cultures, which are the foundation for the Chinese ecological civilization to stand tall and firm in the world. The idea that man must seek harmony with nature is the main body of the traditional Chinese culture and the characteristic of Chinese ecological civilization. When building an ecological civilization, the Chinese people must protect, inherit and innovate the cultural genes of our ancestors. We must know who we are and where we are going to. Based on this, we will consolidate our foundation. Third, we will cultivate and develop core socialist values. This is the basic project of integrating ecological civilization into cultural advancement. General Secretary Xi Jinping pointed out: "The core values carry the spiritual pursuit of a nation and a country and are the most lasting and deepest power" [45] Core values are highly concise and concentrated expression of the socialist core value system. Among them, civility, harmony, equality and justice must also contain the civility of ecological civilization, harmony between man and nature, equality between man and nature, and environmental justice shared by contemporary and future generations. We should build ecological civilization and foster ecological culture in the process of strengthening the building of socialist core value system and core values.

We will insist on integrating ecological progress into social development. First, we must focus on solving the most pressing, most immediate issues that concern the people the most. This is the focus of integrating ecological civilization into social advancement. At present, we need to pay special attention to the prevention and control of water, air and soil pollution. This means that we should focus on the prevention and control of water pollution in river basins and regions, air pollution in key industries and areas especially particulate matter pollution, and the comprehensive treatment of heavy metal pollution and soil pollution. We will concentrate on solving outstanding environmental problems that harm people's interests, such as fine particulate matter pollution, drinking water pollution, soil pollution, heavy metal pollution and pollution by chemicals. Second, we need to pay close attention to the rural ecological security in urbanization. This is the focus of integrating ecological civilization into social progress. General Secretary Xi Jinping pointed out: "Rural civilization is a main part in the history of Chinese civilization, villages are the carrier of this civilization, and farming-reading culture is our soft power. The integrated development of urban and rural areas enables us to preserve the original style of villages. We will be cautious at cutting down trees, filling lakes and demolishing houses. We will try our best to improve the living conditions of residents with original village forms unchanged" [46]. Third, we will promote the modernization of the ecological environment governance system. This is the starting point of integrating ecological civilization into social advancement. The goal of comprehensive and deepen reform is to improve and develop the system of socialism with Chinese characteristics and to modernize China's system and capacity for governance. The ecological environment governance system is a part of China's system for governance. The key to the modernization of the ecological environment governance system is the building of a governance mode involving multiple entities. It is necessary to break the management mode of absolute leading and one-way promotion by government departments, and further form a new pattern of promoting ecological progress with the government as the leading factor, enterprises as the main body, the market effectively driven and the whole society participating together.

Chapter 2
Lucid Waters and Lush Mountains are Invaluable Assets

Tightening resource constraints, serious environmental pollution and severe ecosystem degradation are what we need to face in the present era. Sustainable development of human society is also confronted with serious challenges. In order to solve ecological and environmental problems and actively respond to the challenges, since the early 1970s, a variety of academic trends of research, social movements, government action and market exploration have emerged in the world with environmental protection, sustainable development and green economy as their themes. Human beings have reflected on the mode of industrial civilization and the traditional extensive development mode, re-examined and located the relationship between man and nature, and explored the path of sustainable development. Promoting ecological progress is the strategic innovation and first practice made by the Chinese communists under such a macro background. The core of its theoretical basis and guiding ideology is the "Two-mountain Theory" proposed by Xi Jinping, General Secretary of the Communist Party of China (CPC). That means "We want to have not only mountains of gold, but also mountains of green. If we must choose between the two, we would rather have the green than the gold. And in any case, green mountains are themselves gold mountains". After ten years of theoretical development and practice examination, the "Two-mountain Theory" has become more and more mature and been incorporated into the Central Document. The "Two-mountain Theory" has become an important guiding thought for China to accelerate ecological conservation and a significant part of governance of the current CPC Central Committee.

1 We Want to Have not Only Mountains of Gold, but also Mountains of Green

"Mountains of green" refers to the quality ecological environment, as well as its associated ecological goods. "Mountains of gold" represents economic incomes, and people's well-being related to income and growth level. Therefore, "mountains

J. Pan, *China's Global Vision for Ecological Civilization*,
https://doi.org/10.1007/978-981-16-4534-1_2

of green" and "mountains of gold" essentially point to the relationship between environmental protection and economic development. How should we view, coordinate and unify the relationship between the two? General Secretary Xi Jinping pointed out, "We pursue harmony between man and nature, and harmony between economy and society. In other words, we want to have not only mountains of gold, but also mountains of green". This is a summary of man and nature, and economy and society. It points out that the development of human civilization should follow "harmony between man and nature, and harmony between economy and society". At the same time, it also explains the dialectical relationship between economic development and ecological environment protection in the process of comprehensively deepening reform. We need to develop the economy as well as protect the ecological environment. In the present era, the principal contradiction facing Chinese society has evolved. What we now face is the contradiction between unbalanced and inadequate development and the people's ever-growing needs for a better life. The modernization that we pursue is one characterized by harmonious coexistence between man and nature. In addition to creating more material and cultural wealth to meet people's ever-increasing needs for a better life, we need also to provide more ecological goods of better quality to meet people's ever-growing demands for a beautiful environment.

Development is of paramount importance. Since the beginning of human society, the eternal theme has been realizing harmonious coexistence between economic and social development and environmental protection while maintaining the dynamic balance between human development and natural ecosystem as a whole. It is only in different stages of the development and progress of human society that the principal and secondary contradictions are manifested in different forms, and the principal and secondary aspects of the contradictions have different forms of mutual transformation. Marx pointed out: "The first premise of human history is undoubtedly the existence of living individuals. ... Any historical record should start from these natural foundations and the changes that have taken place in the course of history due to people's activities". He also said: "Only in society is nature the foundation of man's own existence. Only in society, the natural existence of man is his existence". Therefore, both mountains of green and mountains of gold are important factors for economic and social development of human beings, and should not be biased.

We usually say that food, clothing, shelter and means of travel are the four basic needs of human survival. How to meet the basic needs of human survival and how to solve this problem well depend on what social products are produced, and how they are produced, circulated and distributed. Economic development is the goal that human society has been striving for. Therefore, development comes first, but understanding what development is and how to realize it is the key to our work and right actions. There is no treasure bowl or cash cow in the world that can conjure up gold, silver and precious things out of thin air, and tables that can conjure up delicacies only appear in fairy tales. Where does wealth come from? As the ancients said, if one is industrious, one will not be short of supplies. Marx's historical materialism has always held that the creation and accumulation of wealth in society come from nature and through hard work of human beings. Since the beginning of human society, in the course of long-term struggle for production, production practices and scientific experiments, human beings have been continuously knowing, utilizing and transforming nature so that nature can provide benefits to human beings, thus promoting the continuous progress of human society.

In different developmental periods of human society, there have been different pursuits and needs. However, no matter what stage of development, we should not waver in our commitment to development. Development is of paramount importance. This is especially true for the economic and social development of tortured modern China. Due to poverty and backwardness, all Chinese people know clearly of the hardships endured by the Chinese nation in modern times so that they are very eager for development. Since the reform and opening up, industrialization and urbanization have progressed by leaps and bounds, and the development paradigm of industrial civilization has become the mainstream. Economic and social development, comprehensive national strength and international influence have made historic progress. Chinese people have created an amazing "Chinese Miracle" in the history of world's economic development with their diligence, tenacity and wisdom. Taking the statistics of the National Bureau of Statistics on the 65th Anniversary of the Founding of the People's Republic of China as an example, from 1953 to 2013, China's gross domestic product (GDP) increased 122 times at comparable prices, with an average annual growth rate of 8.2%. In 1952, the GDP was only 67.9 billion yuan, but in 1978 it rose to 364.5 billion yuan, ranking tenth in the world. Since the reform and opening up, The GDP has increased by 9.8% annually, and the growth rate and the duration of high-speed growth have exceeded those of Japan and South Korea during their economic take-off. The GDP jumped to the new level continuously, exceeding 1 trillion yuan in 1986; 2 trillion, in 1991, and 10 trillion, in 2001. It reached 40 trillion yuan in 2010, surpassing Japan to become the world's second-largest economy. In 2013, it reached 568.845 billion yuan, accounting for 12.3% of global GDP. China's per capita GDP increased from 119 yuan in 1952 to 41,908 yuan (about US$6,767) in 2013. According to the World Bank's classification criteria, China has moved from a low-income country to an upper middle-income country. In this process, frankly speaking, it was not rare to find that mountains were damaged for mining, and ponds were filled for spaces to build factories on. It was quite phenomenal to find people pursuing the economic growth with the focus on "short cycle, less investment and quick effect" projects and rushing into "high pollution, high energy consumption and low output" projects. The unbalanced, uncoordinated and unsustainable contradictions that accompany the rapid economic growth were still prominent. On the whole, however, this period was exactly what General Secretary Xi Jinping called "We want to have not only mountains of gold, but also mountains of green". In response, Xi Jinping pointed out: "China's eco-environmental conflicts are formed gradually. The eco-environment does not deteriorate in a single day, but we cannot make it worse and worse. Communists should have such a mind and will".

As China enter a new norm in economic and social development, to strive for green, low-carbon and circular developments has become the mainstream voice and practice direction of economic and social development. However, whether it is green, low carbon, circular or ecological, the aim is for development. China today still gives top priority to development in governing and rejuvenating the country by the Party. Just as the General Secretary Xi Jinping pointed out: "As long as there is no fundamental change in the general trend at home and abroad, we cannot and should not change our focus on economic development. This is the fundamental requirement to commit to the Party's basic line of 100 years and to solve all problems in China today." At the same time, we need to pay attention to the fact that, as the fundamental source of invaluable assets, clear waters and green mountains serve to be the foundation for sustainable development of mankind, which must be resolutely guarded. In his speech at the meeting of the Standing Committee of the Political Bureau of 18th CPC Central Committee, General Secretary Xi Jinping said: "If it is still extensive development, even if the goal of doubling the GDP is achieved, what kind of pollution will it be? I'm afraid it will do damage to the resources and environment." We are warned to control our own greed, to be awed by nature all the time, to know how to act according to the laws of nature, and at the same time to clarify the deep-seated factors that cause the contradiction between clear waters and green mountains and mountains of gold and silver, which lie in the wrong thinking mode and linear development mode of single-dimensional, subject-object opposition.

Development must be sustainable development that follows the laws of nature. This is an inevitable conclusion we have drawn from numerous experiences and lessons and an inevitable choice for our country to further deepen reform. We must consider the margin of industrialization and economic growth. If we blindly exploit or destructively utilize natural resources, we will lose the value of choice and existence value of those natural resources that have not been recognized by the market, and finally the development of human beings will come to an end.

At present, we are devoting to ecological conservation. Our goal is to realize the human beings' developing in harmony with nature. What we need is protection in development, which is not to return to the original working and living style, nor is it to continue the development mode of industrial civilization in pursuit of profit maximization. Yet it is to achieve the maximization of economic, ecological and social values including ecological values. We must observe the laws of nature, respect nature, follow its way and protect it. We must build a civilized society featuring increased production, higher living standards, and healthy ecosystems based on the resources and environmental carrying capacity, and seek sustainable development. "The rise or fall of a civilization is closely tied to its relationship with nature". It requires us to adhere to "green development" in accordance with "five-sphere integrated plan" and the "four-pronged comprehensive strategy". We should give priority to resource conservation and environmental protection. We should adhere to the principle of protection in development and development in protection, and realize economic and social development and ecological environment protection advance together. While enjoying the benefits brought by economic development, the people can feel the improvement of living and working environment, thus comprehensively improve the index of happiness.

General Secretary Xi Jinping's assertion that "We want to have not only mountains of gold, but also mountains of green" embodies the development philosophy of the Chinese Communists, makes further understanding of implications of development, and indicates the reflection of the old extensive development mode, which has confirmed China's choice to develop eco-friendly growth models. The assertion is full of the philosophy of ecological method, applies innovative thinking methods of the theory that everything has two aspects and the theory that studies abstract organization of phenomena, independent of their substance of Marxist philosophy, makes it clear that development is the first priority, and protection is an important criterion to judge whether the development route is correct or not, which requires us to usher in a new era of ecological civilization with the vision of sustainable development.

2 We Prefer Clear Waters and Green Mountains to Mountains of Gold and Silver

When the unity of opposites of economic development and environment conservation are presented to people at the same time, General Secretary Xi Jinping sharply pointed out: "Under the circumstance that we cannot have a cake and eat it, we must comprehend opportunity cost, choose wisely, learn to discard the useless, and do things must be done and things must not be done. We will unswervingly implement the Scientific Outlook on Development, and build a resource-efficient and environment-friendly society achieving harmony between human and nature." He stressed that "the economy should develop, but not at the cost of damaging the environment." "We prefer clear waters and green mountains to gold and silver mines." This assertion clearly shows that once there is conflict and contradiction between economic growth and ecological conservation, we should put ecological conservation first without hesitation, and we should be in no way follow the old path of pursuing for invaluable assets by damaging the environment. "We must firmly keep in mind the concept of ecological red line. In the issue of ecological conservation, we must not cross the line, otherwise comes the punishment." These words fully demonstrate the Party Central Committee's firm will and determination to strengthen environmental protection, and also show the further understanding to the laws of socialism with Chinese characteristics.

In our living space, the social system, the economic system and the natural system are coupled into a compound ecosystem through human activities, that is to say, human society ecosystem. In this system, all elements are interdependent, mutually restrictive and interactive with each other. Human economic activities are limited by the capacity of natural ecosystems, yet the results of human economic activities—social systems and economic systems create reaction on natural ecosystems. Being independent and open, each system has its own operating rules as well as is affected and restricted by other systems. Only when each system adapts to each other and the input and output are generally equivalent, the entire complex ecosystem can reach a balance, and can it continue to circulate stably and continuously. In the conventional analysis of economic growth, although environmental factors have not been explicitly included in the analysis of input and output, environmental constraints on the economic system always exist. Especially with the economic growth, when the rate of resource consumption exceeds the rate of resource renewal, and the discharge of waste exceeds the self-purification capacity of the environment, environmental problems gradually become acute and prominent. When technological progress still cannot ensure that economic development is within the load range that the environment can bear, the ability of the environment to provide resources is no longer the degradation expressed by the environmental Kuznets curve, but the complete loss of the ability to produce and reproduce. At that time, the balance of the ecosystem will be destroyed, and even if great efforts are made to repair it, it will be difficult to restore the original ecology, which becomes the so-called "environmental irreversibility".

In this regard, our forefathers have rich wisdom of ecology. Chinese ancient philosophers have expounded the philosophy of ecological system theory that "Heaven and earth were born at the same time I was, and the ten thousand things are one with me." *Yi Zhou Shu* has also recorded: "In this way, the land would keep its soil fertility, all things on the earth would maximize their roles, everybody would pay their dues, and heaven would follow the rules of nature, which works to make fortunes." Only by achieving harmony with resources and environment can human beings evolve and prosper. As crucial security for human's survival and development, ecological environment is pivotal to a country's or a region's comprehensive competitiveness. A large number of examples have proved that we must respect nature, be in awe of it and protect it if we want the sound development of economy and society. Any act against nature or trying to override the laws of nature is inadvisable, otherwise nature will take its revenge.

In the new era, great progress in productivity and major breakthroughs in production technology have made the consumption rate of natural resources greatly exceed its own repairing rate, while the large amount of waste and poisonous and harmful substances produced by human activities exceed the absorptive capacity of environment, which is what we now call the eco-environmental capacity. Given this, the living environment of human beings is deteriorating. Fresh air is polluted, clean water sources are polluted, toxic and harmful agricultural products produced by heavy metal contaminated soil are harming people's health. The climate is warming, resources are drying up, ecology is deteriorating, the gap between rich and poor is widening, and the urban boundary is expanding indefinitely. Natural disasters such as typhoons, floods, droughts and earthquakes are increasing year after year under the influence of human beings, and the development of human society is facing unsustainable challenges. Judging from the trend of ecosystem deterioration, there are not only the common non-biological toxic and harmful substances that cause changes in the composition of the atmosphere, water and soil, but also global hazards such as ozone hole and global climate change caused by greenhouse gas emissions. In addition, there are also the destructive and wasteful exploitation and use of non-renewable resources such as coal and oil, as well as the destruction of biological resources such as species extinction and biodiversity reduction, etc., on this very note, the interplayed biological and non-biological destruction jointly deteriorate the ecological environment, all of which drive the earth's ecological environment close to the limit of human living environment.

China is a large country with a population of more than 1.3 million. It's not feasible for us to advance modernization by following the old path of the US and Europe. The relatively insufficient energy resources and the weak eco-environmental carrying capacity have become the basic national conditions of our country. Environmental problems in developed countries in the past 100 or 200 years have been manifested in China's rapid development over the last 30 years, presenting obvious structural, compressive and composite characteristics. Some old environmental problems await solutions while the new ones have followed. Sustained and sound social development will not be achieved if we continue to follow the old path, consume resources immoderately and pollute the environment at all cost. General Secretary Xi Jinping pointed out, when people live upon the mountain, they will soon be left with nothing to eat. The ecological destruction here will exert an impact on the overall situation of the country. Ecological construction will be delayed until it's polluted and destroyed. We must not succumb and be tolerant to any acts of polluting and damaging environment. Therefore, we, as human beings, must respect nature, never stop revering nature and observe the laws of nature. As an outcome and a part of nature, the existence and development of human beings depend on the natural environments. To protect nature is to protect human beings themselves. Only in this way can the development of society and production be fundamentally guaranteed.

For a long period in the past, we believed that there was a mutually independent and even opposite relationship between environmental protection and wealth growth. It's a misunderstanding. Meanwhile, with inadequate assessment system and wrong view of achievement, GDP was blindly pursued in the process of development in some places. They simply regarded GDP as an important criterion and exchanged lucid waters and lush mountains for invaluable assets. As a result, resources damage and waste were still serious. Environment impact assessment (EIA) became a formality. Non-standard polluting enterprises have been operated. Illegal projects have been approved. Major pollution incidents happened frequently. That mode of development, which simply focuses on GDP, benefits the few but greatly harm fundamental interests of the majority of our people. Many of the damages caused by ecological environment destruction were discovered or deteriorated after a long period of time. Pollutants, with diffusivity, may spread through the atmosphere, rivers and soil. Its serious consequences may not be detected for a short while. People's low take of pollutants does not react to the human body instantly, however, when it accumulates gradually, the side effects on the body will trickle in. Therefore, we should stop destroying lucid waters and lush mountains in exchange for invaluable assets without thought to the future. General Secretary Xi Jinping repeatedly stressed that we must not get the development of economy for a period of time at the cost of the ecological environment.

As a significant factor of production, to destroy lucid waters and lush mountains is to destroy the ecological environment, then lose the basic conditions for economic development, and the foundations of invaluable assets. Where there is life, there is hope; where there are lucid waters and lush mountains, there is the foundation for sustainable development, which can turn lucid waters and lush mountains, or ecological environment, into productivity and turn economic superiority into economic advantage. Lucid waters and lush mountains can bring invaluable assets, but invaluable assets cannot buy lucid waters and lush mountains. Where there are no lucid waters and lush mountains, there are no invaluable assets. While when they two contradict to each other, we prefer lucid waters and lush mountains than invaluable assets. We must hold on to the bottom line of environment. Only when we attach great importance to the ecological environment as an essential factor of productivity, observe the laws of nature, and protect and utilize the ecological environment can we better develop productivity and realize the harmony between human beings and nature at a higher level.

General Secretary Xi Jinping's theory of "preferring lucid waters and lush mountains than invaluable assets" essentially stresses giving priority to ecological progress and environmental protection. It emphasizes that sustainable development should be realized on the basis of "preserving lucid waters and lush mountains". It's a re-sublimation of the thought of "we want to have not only lucid waters and lush mountains, but also invaluable assets" and a unity of the theory of Marxist philosophy that everything has two aspects and that everything has its key points. It runs through the basic philosophy of "pursuing harmonious development between man and nature, respecting nature, following its ways, and protecting it." It requires us to follow the rules of economic development as well as the laws of nature development, giving priority to the latter. When the two conflict to each other, the right choice must be made, that is, economic development is premised on following the laws of natural development.

3 Lucid Waters and Lush Mountains are Invaluable Asset

Nowadays, the importance of environmental protection has been commonly recognized. However, how to protect it? One thought is negative and passive, that is, to slow down the pace of transforming nature. In this way, we would not go back to the original wildwood to live a natural and original ecological life by eating raw animal flesh, but enjoy certain achievements of modern industrial civilization. However, under the circumstances that China's entered a new normal in economic development and the international competition of economic globalization has intensified, though it's also an option, it could never be our choice. What we should do is to actively protect environment in the process of development, relying on scientific and technological methods and all-round reform and innovation to achieve a higher level of protection. General Secretary Xi Jinping proposed that "lucid waters and lush mountains are invaluable assets" and that lucid waters and lush mountains should be turned into invaluable assets.

There were two misconceptions existed. One is that development will inevitably lead to environmental destruction, which forms the ideological basis of "GDP-first" theory; the other is that paying great attention to protection must be achieved at the cost of sacrificing or even giving up development, turning into an excuse for lacking drive. General Secretary Xi Jinping's proposal of "lucid waters and lush mountains are invaluable assets" pointed out the transformation of green development mode and defined the ecological thinking mode. It has important theory significance and practical guiding value for correcting the above misunderstandings. The proposal of "lucid waters and lush mountains are invaluable assets" profoundly reveals the rules of ecological value realizing and increasing in ecological progress and further develops and improves the Marxism's value theory.

The classical Marxist labor theory of value explains the relationship between human beings. The complex social interest relationship is essentially a value relation. As far as the relationship between man and nature is concerned, whether human beings are the objects of natural products or the subjects of understanding, developing and utilizing nature, it's also embodied as a value relationship, the foundation of human social relationship and the core of the whole ecosystem. The classical Marxism theory attaches importance to the value of natural resources. The natural and eco-environment resources, represented by lucid waters and lush mountains, have the value cycle within the self, which play a significant role in maintaining the stability and balance of the ecological system and creating living conditions for human beings. In fact, natural resources not only produce economic products, but also supply the oxygen we breathe and clean water sources, dispose of waste, beautify the environment and improve the happiness of people living in it. It's clear that natural resources have not only economic value, but ecological value and social value. Currently, we are making great efforts to promote ecological progress. An important aspect of that is to realize the transformation and renewal of ecological concepts, continuously deepen the understanding of the essence of the relationship between man and nature, attach great importance to and restore the value and wealth of ecology, so as to lay a solid ideological foundation for ecological progress.

Value orientation influences the development direction. In different phases of development, people assign different values to natural resources, resulting in great differences in behavior choices corresponding to value orientation. Since the Industrial Revolution, the one-sided view of the value of natural resources and the mechanical interpretation of Marxism have led to the artificial narrowing of the extension of the concept of resource value. We destroyed lucid waters and lush mountains for our needs of mountains of gold and silver. As a result, GDP has increased, but a serious of problems such as resource shortage, environmental pollution and ecological balance disturbance have been brought about. Over-exploitation and even plunder of natural value have resulted in serious overdraft of natural value and triggered global ecological crisis. Natural value has gone towards debt.

General Secretary Xi Jinping pointed out that "lucid waters and lush mountains are invaluable assets", which fully embodies the value philosophy of respecting nature, thinking highly of the value of resources and seeking the harmonious development between man and nature. It's the heritage and development of Marxist core value theory and the oriental wisdom of contemporary China.

From the perspective of view of development, to realize lucid waters and lush mountains is to obtain invaluable assets, whose essence is to realize ecology-based economy development and economy-based ecology development. Poverty is not an ecological civilization, and development cannot destroy the environment. On the one hand, in order to protect ecology and restore the environment, we can no longer pursue economic growth at the cost of massive consumption of resources and environmental destruction. We should instead guide the development of eco-driven and eco-friendly industries, that is, ecology-based economy development. On the other hand, we should transform the high-quality ecological environment into residents' monetary income. We can endow the high-quality ecological environment with a reasonable market price according to the scarcity of resource to respect and reflect the ecological value of the environment, and further carry out the compensation system for use and trading, that is, economy-based ecology development. The development of ecology-based economy development requires us to set up correct values, take structural adjustment as the starting point, transfer mode, adjust structure, change direction and improve quality; the advancement of economy-based ecology development needs us to promote the institutionalization of property rights to implement the compensation system for use and trading of natural resources property rights such as water rights, mining rights, forest rights, fishing rights, energy rights, environmental resources property rights such as ecological rights and discharge rights and climate resources poverty rights such as carbon credit and carbon sink.

Nowadays, the ecological environment is increasingly becoming an important source and safeguard for the development of productive forces. When General Secretary Xi Jinping attended the Guizhou delegation of the 19th CPC National Congress, he pointed out that socialism with Chinese characteristics has crossed the threshold into a new era, so we must provide more ecological goods in accordance with the requirements of the new era to meet people's ever-growing needs in many areas and better promote well-rounded human development and common prosperity for everyone. Marxism's Theory of the Productive Forces has also told us that productive forces include not only human beings as workers and their creativity, but also the external ecological environment. For example, we will insist on the economical and intensive use of resources, rely on the advantages of ecological environment to develop green industries, and make full use of good ecological environment to attract high-tech talents and modern industries with high-tech as the core. As such, the beautiful ecological environment will become an important "natural capital", and bring more development opportunities. The development potential will be improved, new impetus will be formed, ecological dividends will be generated, and lucid waters and lush mountains will continue to bring invaluable assets. Take Zhejiang Province as an example. It has effectively acted on the understanding of "Two-mountain Theory", and put into practice the strategy of "making full use of eight advantages and implementing eight major measures" for a decade. By combining environmental protection and the promotion of ecological economy to resolve the contradiction between the two, it has coordinated environmental protection with the transformation and upgrading of enterprises, the change of government management mode, and the promotion of the system of resource property rights. It has proved that lucid waters and lush mountains can be turned into invaluable assets, and environmental protection and wealth growth have entered a virtuous circle of mutual enforcement. Thus, Zhejiang has achieved higher-quality and sustainable economic growth, dealt with many difficult problems that cannot be solved in the traditional industrial economic system, and created a new eco-economic model featuring benign interaction between natural capital growth and environmental improvement.

We will make innovations based on the reality, capitalize on the ecological and economic advantages to develop green industry and beautiful economy, and improve the supply of ecological goods. In terms of the ways of changing the lucid waters and lush mountains into invaluable assets, there is no universally applicable model, and we should make a concrete analysis of a concrete problem. With local conditions in mind, all places must make explorations actively, have the courage to pursue innovation and development, and adhere to a characteristic development model. General Secretary Xi Jinping pointed out that industrialization does not mean that we should operate industries everywhere, but develop industries, agriculture, exploitation and protection according to local conditions. For this reason, General Secretary Xi Jinping stressed: "We must innovate our development ideas and give full play to our late-comer advantages. In light of local conditions, we should choose and develop different industries. Thus, we can make full use of lucid waters and lush mountains to generate economic and social benefits, effectively achieve the simultaneous improvement of economic, social and ecological benefits, and realize the unity of people's wealth and ecological beauty."

China has not completed its industrialization process, which is different from post-industrialized countries that have completed their economic transformation and from the countries in the early stage of industrialization that have the price advantage of factors of production. China is not only facing the requirements of speeding up development, realizing industrialization and avoiding falling into the middle-income trap, but also facing five aspects of challenges: shortage of resources, environmental pollution, ecological degradation, people's urgent desire for a better life and international competition. In the process of modernization, as a responsible large developing country with a population of more than one billion, China must bear in mind the fundamental change of development philosophy and model. General Secretary Xi Jinping further pointed out: "The economic growth which solely relies on the stimulant policies and the government's large-scale direct intervention will only treat the symptoms but not the root causes. The growth based on a large amount of resource consumption and environmental pollution can hardly sustain. We should improve the quality and efficiency of economic growth and avoid solely focusing on GDP growth rate." Therefore, changing lucid waters and lush mountains into invaluable assets, realizing the transformation of green development and adjusting economic structure are the inevitable choices to break through the bottleneck of resources and environment and realize sustainable development. General Secretary Xi Jinping stressed that we must be completely changed our ideas. Lucid waters and lush mountains and invaluable assets are by no means opposites. The key lies in people and thinking. We must meet a prerequisite for giving full play to people's subjective initiative, that is, we must comply with rules and resolutely reject the mechanistic views of denying, conquering and transforming nature. Promoting productive forces to enter a new stage of development is an urgent need for the survival and development of the present and future generations, and a strategic need for the liberation and expansion of productive forces. Only by making mountains and rivers green again can we bring wealth to the people. Lucid waters and lush mountains are invaluable assets. General Secretary Xi Jinping further pointed out in the report delivered at the 19th CPC National Congress, "Building an ecological civilization is vital to sustain the Chinese nation's development. We must realize that lucid waters and lush mountains are invaluable assets and act on this understanding. We must pursue a model of sustainable development featuring increased production, higher living standards, and healthy ecosystems. We must continue the Beautiful China initiative to create sound working and living environments for our people and play our part in ensuring global ecological security. "

General Secretary Xi Jinping's theory of "Lucid waters and lush mountains are invaluable assets" is a dialectical unity of the theories of "We want to have not only mountains of gold, but also mountains of green" and "If we must choose between mountains of gold and mountains of green, we would rather have the green than the gold". It is an innovation in development theory, reflects a new height in the development of Marxist theory, greatly enriches and expands the Marxist Theory of Development, and represents a major innovation in the values of socialism with Chinese characteristics.

To sum up, General Secretary Xi Jinping's "Two-mountain Theory" stresses that "we want to have not only mountains of gold, but also mountains of green, and if we must choose between the two, we would rather have the green than the gold". The theory is based on the reality and the issues that people are more concerned about, and is gradually summarized and developed from the growth and practice of local areas. It expresses profound and systematic thoughts, vividly summarizes the unity of opposites of development and protection, and reflects the transformation and promotion of scientific view of development, ecological view, values and view of achievement. "We want to have not only mountains of gold, but also mountains of green" emphasize that we must take a holistic approach to develop economy and protect environment, and continue to pursue development as the Party's top priority in governance. "If we must choose between the two, we would rather have the green than the gold" means that ecological progress and environmental protection should be high on the agenda. "Lucid waters and lush mountains are invaluable assets" reveals the real value of ecological environment and reflects the return of people's understanding of the natural ecological value.

In addition, the "Two-mountain Theory" objectively analyzes the principal and secondary contradictions and the principal and secondary aspects of the contradictions between development and protection, answers the complex question of the relationship between human and nature, and adheres to the unity of "theory that everything has two aspects" and "theory that everything has its key points". It is the inheritance and development of the ancient Chinese civilization's "harmony between man and nature" thought, the summary of China's practical experience in economic development and the development of the theory of sustainable development. The "Two-mountain Theory" is innovative thinking and revolutionary achievement of dialectical materialism in integrating theory with practice and guiding practice with theory in governing the country. It's negation of the negation of "growth-pollution-protective growth",and an innovation in the ecological progress theory of negation of the negation of “environmental protection-slowing down growth-better growth". It is an important symbol of the initial formation of Xi Jinping’s ecological philosophy and a new development of Marxism in contemporary China, and shows the building direction of a Beautiful China.

Chapter 3
Environment is People's Livelihood, and Environment is Productivity

Ecological environment is an important part of the comprehensive competitiveness of a country and a region, and the guarantee and embodiment of people's basic living conditions and quality of life. Environmental protection underlies development, people's livelihood and their basic development opportunities, abilities and rights and interests. Environmental protection will benefit the present and future generations. With the progress of society and the improvement of people's living standards, people's demands and requirements for a good ecological environment are increasing. The importance of ecological environment in the National Happiness Index is constantly highlighted. Since the 18th National Congress of the Communist Party of China, the CPC Central Committee and the State Council have attached great importance to environmental protection as an important livelihood issue and development strategy. General Secretary Xi Jinping pointed out: "A sound ecological environment represents the fairest public good and the most universal form of public welfare" [1]. He repeatedly stressed: "The environment is people's livelihood, the green hills are beautiful, and the blue sky also represents happiness" [2]. "We must protect the ecological environment like protecting our eyes and treat the ecological environment like our lives" [3]. Taking environmental protection as an important livelihood issue profoundly reveals the essential connotation and ultimate goal of environmental protection, embodies General Secretary Xi Jinping's deep love of the people with the fundamental interests of the people as the center, represents a concrete manifestation of people-oriented governing concept, and enriches and develops the basic connotation of people's livelihood.

J. Pan, *China's Global Vision for Ecological Civilization*,
https://doi.org/10.1007/978-981-16-4534-1_3

1 A Sound Ecological Environment Represents the Most Universal Form of Public Welfare

Environment is the basic condition of people's life and the basic factor of social production, and it is in the fundamental interests of the Chinese people. If the environment is well protected, all citizens will benefit. Otherwise, the society will suffer. The condition and quality of the environment directly affect people's living conditions, thus affecting the level of social development and ultimately determining the success or failure of civilization. General Secretary Xi Jinping pointed out: "Our people love life and hope for better education, more stable work, more satisfactory income, more reliable social security, higher level of medical and health services, more comfortable living conditions and more beautiful environment. They also expect children to grow up, work and live better. The aspirations of the people to live a better life will always be the focus of our efforts" [4]. "Putting the building of an ecological civilization in a more prominent position is where public demand lies" [5]. People are not dissatisfied with the growth rate of GDP, but more dissatisfied with the poor ecological environment. We must find out what our people want. On the one hand, in the low stage of social development, people struggled for survival and strove for food and clothing, so getting rid of poverty and backwardness was the main focus of social governance. On the other hand, we were limited by the labor force, so the exploitation and utilization speed of the environment was far lower than the self-repair speed of the ecological environment, so the environmental problem as an attribute of people's livelihood was not highlighted. However, modern science and technology develop rapidly in modern times. On the one hand, the development of modern science and technology makes people's production and life more convenient, thus promoting the progress of human society. On the other hand, due to the overexploitation and overdraft of natural resources, the consumption exceeds the regeneration rate, and the discharge of pollutants exceeds the environmental capacity, so environmental problems are constantly emerging and seriously threaten people's life and health.

Environmental protection relates to both development and people's livelihood. Everyone needs to breathe clean air, and drink clean and fresh water. Non-polluted soil is the most basic condition for food production. Therefore, ecological environment, as a special public product, is more important than others. However, the quality maintenance of air, water and soil has strong externalities. To protect them from pollution, we may conflict with the economic interests of some small collectives, resulting in the tragedy of commons. Small collectives only care about their own interests, think that the overall ecological environment protection is a matter of the society and country, and everyone pursues development in a beggar-thy-neighbor way. In this case, the result is that the living environment of the whole country and even mankind will be impacted or even completely destroyed. Today, the people of China find it difficult to drink clean water, breath fresh air and eat safe food, which is the simplest wish of building an ecological civilization. This reflects the theoretical dilemma and practical difficulty.

As early as 2007, when the term "ecological civilization" was written into the report for the first time at the 17th National Congress of the Communist Party of China, all sectors of society interpreted it as a new civilization form of "the win-win of increased production, healthy ecosystems and happy life". The 18th National Congress of the Communist Party of China further incorporated the building of an ecological civilization into the national development strategy. General Secretary Xi Jinping emphasized: "Economic development and increased GDP are not the only things we pursue. We must also pay attention to the indicators of social progress and civilization prosperity, especially indicators of culture, resources and environment. We must not only work hard for today's development, but also be responsible for tomorrow's prosperity, and lay a sound foundation for future development and provide resources and environment that can be utilized sustainably" [6]. In August 2016, when General Secretary Xi Jinping presided over the National Health and Fitness Conference, he pointed out: "Healthy ecosystems underlie human survival and health. We must implement the strictest ecological environmental protection system in accordance with green development. We will establish and improve environment and health monitoring, investigation and risk assessment systems, focus on the prevention and control of air, soil and water pollution, and accelerate afforestation. Therefore, we can effectively solve the outstanding problems affecting people's health [7]. General Secretary Xi Jinping raised environmental protection to the level of people's livelihood and major strategic issues, and attached great importance to it. This reflects his deep love of the people and the Party's purpose, and represents the core essence of the theory of the building of an ecological civilization in Xi Jinping Thought on Socialism with Chinese Characteristics for a New Era.

In fact, in terms of economic development and environmental protection, we must take the fundamental interests of all our people as the starting point and ultimate purpose. In the concept system of classical Welfare economics, public products do not include air and water that can be obtained without human labor, but include medical services, educational resources and employment opportunities that entail social efforts. A healthy ecological environment means clean air, clean water, safe food, rich products and beautiful landscape. If it is damaged, both the rich and the poor are not spared, so the ecological environment is obviously universal and fair. As ecological environment problems become increasingly severe and exerts greater impact on social life, the attribute of ecological environment as a social product is more and more obvious. "The fairest public good " emphasizes not only the importance of governance results but also the primary responsibility of governance process. It is the basic obligation and important function of governments at all levels to shoulder the primary responsibility of environmental governance. It is necessary to plan and guide ahead so as to promote greening in the whole process of economic and social governance. Only in this way can we learn the lessons from the early-developing countries, avoid the costly and disorderly environmental protection, and turn passivity into initiative. With the goal of providing a happier life for the people in mind, General Secretary Xi Jinping regards a sound ecological environment as the fairest public good and the most universal form of public welfare. This is the perfection, enrichment and development of our Party's people's livelihood thought under the new historical conditions. People's demands for clean water, fresh air, safe food and beautiful environment are getting stronger and stronger, and ecological environment protection shouldn't be promoted slowly as it cannot wait.

At present, China's ecological progress has made great progress and positive results. However, we are facing growing constraints on resources and severe ecosystem degradation, and various environmental pollution incidents such as air pollution, water pollution and soil pollution occur frequently, which affect people's well-being and threat people's basic livelihood. For example, in 2010, due to a series of explosions occurred in the Petrochemical Plant in Jilin City, benzene pollutants flowed into the river and caused water pollution, which disrupted the normal water use of residents on both sides of the river. In that year, a sewage leakage from the Zijinshan Copper Mine caused serious pollution in nearby rivers. From the perspective of ecological conservation in the process of integrated development of urban and rural areas, due to the lack of attention to environmental protection, the environmental protection infrastructure of farmers' concentrated areas lags behind, and a large number of chemical fertilizers and pesticides are applied in farmland. As such, ecosystems in rural areas are damaged more seriously, agriculture and ecological environment are impacted, serious water pollution and eutrophication are caused, and there are excessive residues of toxic and harmful substances in agriculture, livestock and aquatic products. Food is unsafe and endangers people's health, which pose threats to the basic guarantee of people's livelihood.

Besides, the issue of ecological environment which affects the people's life quality and basic livelihood and well-being, has become a major political issue underpinning social harmony and stability. At present, environmental pollution has resulted into social injustice and induced contradictions and conflicts. If this issue is not handled properly, it will undermine social stability, weaken the authority and credibility of the government, and offset the achievements of reform and opening up and economic and social progress. General Secretary Xi Jinping pointed out: "We must protect the ecological environment in a long-term, holistic, and consolidated way. We should not try to save a little but lose a lot, attend to one thing and lose another, live beyond our means and seek quick success and instant benefits".[55] He emphasized that "ecological environment protection will benefit the country, the people and future generations. We should not only talk and not to act" [9]. He put forward that "we must take a scientific approach to lay out production, living and ecological space, take solid steps to protect ecological environment, so that people's life quality will be improved by a sound ecological environment" [10].

Wholeheartedly serving the people is the purpose of the Communist Party of China and the highest value orientation of the Party. Always serving the interests of the people and winning the support of the people is the highest standard to measure whether the Party's lines, principles and policies are correct. Actively responding to the voices of the people and striving to realize and meet the expectations of the people are the starting point and foothold for officials at all levels to think about all problems and the main axis and theme of all work. Strengthening environmental protection and building an ecological civilization are long-term plans concerning people's well-being and the future of the nation. The condition and quality of the environment directly affect people's living conditions. Therefore, protecting the ecological environment means safeguarding people's livelihood, and improving the ecological environment means improving people's livelihood. Ecological protection is an important part of our livelihood work. Governments at all levels must attach great importance to it, keep a clear understanding, and take the interests of the people as the starting point and foothold of our environmental protection work. We must care for the life and health of the people, serve the people, and solve people's most concerned, direct and realistic environmental problems so as to meet the people's growing environmental quality needs. With a sound ecological environment, the people can get a wider space for survival and development, and enjoy the improvement of life quality and the joy of the spirit when witnessing the improvement of material level. As such, the people can step into a higher and better life level.

2 Ecological Environment is the Key Factor in the Building of a Moderately Prosperous Society

"Moderately prosperous society" is a strategic conception put forward by Comrade Deng Xiaoping when he planned China's socio-economic development blueprint in the late 1970s and early 1980s. In 2012, the report of the 18th National Congress of the Communist Party of China officially put forward "building a moderately prosperous society in all respects" for the first time. Since 1970s, the average economic growth rate has been close to double digits, almost three times that of developed countries in the same period. However, the ecological environment, especially the air, water and soil, is seriously polluted, which is becoming a prominent shortcoming of our work of building a moderately prosperous society in all respects. Our practice shows that the high consumption, high input and extensive growth model with resource elements as the core is unsustainable, and the environmental problems that appeared in western developed countries in stages during the industrialization in the past century occur in China in a short time. Under the new normal background, we are facing growing constraints on resources, weaker environmental carrying capacity and severe ecosystem degradation, which has become major factors restricting the sustainable and healthy development of economy and society and affecting people's "sense of fulfillment" and "happiness". For this reason, General Secretary Xi Jinping pointed out that "ecological environment is the key factor in the building of a moderately prosperous society" [11].

Sound ecological environment is a part of a moderately prosperous society in all respects. "In all respects" is a requirement for the intensity, depth and breadth of our work, which is reflected in the comprehensiveness of regions, people and fields our work covers. In a moderately prosperous society in all respects, development of politics, economy, society, culture and ecological environment are push forward. We must continuously improve people's living standards and livelihood, and enable them to enjoy blue sky and green water, healthy body and mind and colorful cultural life. Therefore, in the measurement standards, we should pay attention to the standards of material and livelihood. It is necessary to look at tangible indicators by evaluating the achievements with a scientific and reasonable indicator system, and look at intangible indicators by taking people's satisfaction and dissatisfaction as an important standard. Such comprehensive development can be shared in varied regions and urban and rural areas, is visible, tangible, and can be felt and fully recognized by the people. The moderately prosperous society we want to build is a society in which the fruits of reform and development truly benefit all the people, and is a society with all-round economic, political, cultural, social and ecological development. General Secretary Xi Jinping emphasized: "We must vigorously promote the building of an ecological civilization. We must strengthen comprehensive management measures, address targets and responsibilities, promote clean production, and expand green vegetation. By doing so, we will make the sky bluer, the mountains greener, the water clearer and the ecological environment better" [12].

General Secretary Xi Jinping pointed out: "Enable the people have a greater sense of fulfillment" [13]. According to economics theory, if the ultimate goal of economic and social development is to benefit the people, enable the people to improve their "happiness", and realize the maximization of people's well-being, then ecological progress is more direct and effective in improving people's "happiness". At present, China has reached the decisive stage of building a moderately prosperous society in all respects, environmental problems have been highlighted and environmental protection standards have been raised. China has entered a new normal in economic development, and the disparity between the supply and demand of environmental public services needs to be solved urgently. China's major pollutants and carbon dioxide emissions rank first in the world, and it is in a high emission platform period. The negative impact on groundwater, soil and public health is still rising, and the ecosystem function is very fragile. Although the growth rate of major pollutant emissions has been effectively curbed, the continuous increase in emissions far exceeds the environmental capacity, and the strategic adjustment of industrial structure and energy structure has not yet been completed, so the environmental pressure still exceeds the environmental load. Therefore, during the 13th Five-Year Plan period, it is still necessary to control the emissions of major pollutants. In the next decade, most of the pollutant emissions will exceed the "peak", and the ecosystem is expected to be improved. People's requirements for environmental quality, health and safety are getting higher and higher, and the gap between environmental quality and people's expectations will probably widen further. In this regard, environmental quality may become the biggest shortcoming of building a moderately prosperous society in all respects. How to understand the comprehensive requirements and whether people have "a sense of fulfillment" in this process are related to the effectiveness of this strategic goal. To build a moderately prosperous society in all respects on schedule, it is vital to make efforts to fill the shortcomings of ecological progress, and achieve overall improvement of ecological environment quality.

With an in-depth analysis of people's happiness, we find that people's happiness includes many aspects, such as stable employment and income growth, sound medical care and aged-care social protection, and a good living environment. Ecological environment is closely related to people's life. Meeting the people's new expectations for a good ecological environment is a basic requirement for building a moderately prosperous society in all respects. With the improvement of income level and the expansion of the number of middle-income people, the people have a new understanding of the connotation of happiness. They are increasingly concerned about the environmental problems closely related to life and health, and hope for more blue sky, white clouds, green water and green mountains, and are eager for fresher air and cleaner water sources. The contradiction between the increasing demand for environmental public services and the lagging supply is becoming prominent. Without the protection of ecological environment, development achievements will be greatly reduced, and people's happiness will be hard to improve. If we are to meet the people's demands for ecological environment quality, we must vigorously promote the building of an ecological civilization. We must constantly consolidate the ecological foundation of socio-economic development so that we can build a moderately prosperous society in all respects and make the people's dream of a better and happy life come true.

The building of an ecological environment is related to the building of a moderately prosperous society in all respects. To make up for the shortcomings of ecological environment, it is necessary to raise awareness and take actions. General Secretary Xi Jinping pointed out: "Major ecological restoration projects should be implemented to enhance the production capacity of ecological products. Environmental protection and governance should focus on solving outstanding environmental problems that harm people's health. We must prioritize prevention and comprehensive management. We must strengthen pollution prevention and control of water, air and soil, promote water pollution prevention and control in key river basins and regions, and enhance air pollution control in key industries and key regions" [14]. The Fifth Plenary Session of the Eighteenth CPC Central Committee put forward: "We must build an ecological security barrier, give priority to protection, and let nature restore itself. We must implement ecological protection and restoration projects of landscapes, forests and lakes, carry out large-scale afforestation, improve the natural forest protection system, and take forward 'Blue Bay' remediation action plans" [15]. In this regard, according to the basic requirements of General Secretary Xi Jinping's discourses, we must prioritize prevention, combine prevention with treatment, and concentrate on solving outstanding problems that endanger people's health first. We must use multiple channels to effectively integrate the concepts, principles and goals of ecological progress into all aspects of economic and social development and implement them in various plans and work at all levels. In accordance with the requirements of the 13th Five-Year Plan, we must implement the concept of green development, improve policies and differentiated performance appraisal based on functional zones, and promote the development of various regions based on the main functions. We must give priority to ecological environmental protection, and let nature restore itself. We must implement the ecological protection and restoration project of landscapes, forests and lakes, improve the system stability of the ecological environment, and build an ecological security barrier. We must speed up the transformation of economic development mode and improve various systems of a resource-efficient and environment-friendly society. We will ensure the realization of a moderately prosperous society in all respects, enable the people to enjoy the fruits of reform and opening up, and improve their sense of fulfillment and happiness.

3 Environment Improvement Means Productive Forces Development

Environment is productivity, and productivity itself is the most critical factor of people's livelihood. General Secretary Xi Jinping pointed out: "From the perspective of the history of world development, protecting and improving the environment is tantamount to protecting and improving productivity" [16].

Productivity is people's ability to adapt to nature and transform society. People's understanding of productive forces is constantly deepening [17]. In the process of human society development, the connotation of productive forces has been continuously deepened, and the extension tends to expand. Besides, relations of production and superstructure are constantly changing, and their relationship with productivity is becoming increasingly complex. Productivity development includes not only quantitative expansion, but also frequent changes in quality and structure. On the one hand, ecological environment provides various natural factors among the three basic components of productivity, including land as means of labor, forests and mineral deposits as subject of labor. During the operation of the productivity system, people initiate, adjust and control the transformation of material and energy between human and nature with their own activities. That is to say, the operation of the productivity system is a process in which people constantly ask for wealth from nature to satisfy their own desires and needs by utilizing and transforming nature. In this process, people, as the core component of productivity, and the external ecological environment, as the basic component, interact with each other. In the era of industrial civilization, limited by the understanding of the laws of nature, human beings emphasize the plundering, conquering and defeat of nature. As such, the economic and social development is accompanied by global ecological disasters and crises such as environmental pollution, resource shortage, species extinction and climate warming. In the era of ecological civilization, people have realized the importance of protecting the external ecological environment, and realized that the ecological environment will affect the structure, layout and scale of productive forces, and then affect the operation efficiency and benefits of productive forces.

On the other hand, from the perspective of the effect of productivity on nature, under certain historical conditions, functional upgrading will occur, that is, new functions will appear while the original functions are greatly enhanced. In terms of modern society and future development, it is not difficult to find that in addition to understanding, transforming and utilizing nature, a kind of ability to protect nature (including ecological balance and restoration ability, original ecological protection ability, environmental monitoring ability, pollution prevention and control ability, etc.) has gradually developed within productive forces. Based on this, we should have a new understanding of the connotation of productivity as the core component of Marx's materialism in the new era. Productivity should be people's comprehensive ability to know, transform, utilize and protect nature, and the integration of these four functions constitutes productivity achieving harmony between human and nature. In fact, productivity cannot be separated from nature, and ecological environment is also productivity.

In addition, from the perspective of the relationship between economic system and natural ecosystem, there is unity of opposites between economic system and natural ecosystem. In the eco-economic system composed of economic system and natural ecosystem, the economic system is based on natural ecosystem, and human economic activities are limited by the capacity of natural ecosystem. Both reflect the contradiction between global interests and local interests, long-term interests and immediate interests, and fundamental interests and apparent interests. When the two systems adapt to each other, the ecological and economic balance will be achieved, and the harmonious unity between man and nature will be realized. The compound system will be stable and sustainable, which is a virtuous cycle that we should strive to pursue. When the two systems conflict with each other, it leads to ecological and economic imbalance, the cycle will be destroyed, and the composite system tends to be unstable and unsustainable. The law of coordinated development of eco-economy reminds us that human beings must keep an appropriate level of economic activities in order to realize the coordinated development of eco-economic system [18]. In this respect, we have learnt some important lessons. For example, the Ministry of Supervision has reported ten typical cases of responsibility investigations for destroying ecological environment [19]. Enterprises illegally discharge pollutants, which directly endanger the lives and health of local people and even people in river basins. Another example is that in the rainy season in recent years, many urban areas in China have accumulated so much water that these areas become "sea". One of the most important reasons is that the ecological space is occupied by arbitrarily filling lakes and rivers in urban development.

For a long time, we used to regard people and tools of production as social productivity, and take productivity as people's ability to conquer and transform nature. However, the classical writers of Marxism have never excluded the natural ecological environment from the components of social productivity. In *Das Kapital*, Marx used the concepts of "various social productivity of labor", "all social productivity of labor" and "natural productivity of labor" to clarify the rich connotation of productivity. Marxist concept of productive forces includes not only human labor and creativity, but also nature, which is the basis of human existence and subject of labor. As human beings move towards a new era of ecological civilization, all aspects involved in ecological environment are related to productivity, and ecological environment is increasingly becoming an important productivity. General Secretary Xi Jinping also profoundly pointed out: "Ecological environment is the foundation of economic and social development. Development should be the overall development of the economy and society, the coordinated development in space and the sustainable development in time" [20]. He pointed out: "We should build an ecological system featuring respect for nature and green development. Human beings can use nature and transform nature, but in the final analysis, we are part of nature, and we must care for nature and cannot override nature" [21]. There is no substitute for ecological environment, and it is difficult to survive without it.

In the past, we didn't know enough about the nature of development. We equated development with the growth of GDP, so we simply produced labor products. We didn't realize that ecological products are also one of the necessities of human survival and development, and we didn't understand the natural value and capital of ecological environment. In the process of developing and utilizing natural environment and resources, we failed to correctly handle the relationship between human and nature and limit human activities, thus damaging ecological environment. Practice proves that it is impossible to focus on environmental protection alone without economic development, let alone pursue temporary economic development at the expense of destroying ecological environment. Under the premise of scientific protection and within the scope of environmental carrying capacity, efforts must be made to promote ecology-based economy development and economy-based ecology development, and create a win–win for economic development and environmental protection. It is necessary to establish an overall, long-term view and holistic view, adhere to the basic national policy of saving resources and protecting the environment, protect the ecological environment like eyes, and treat the ecological environment like life. We must ensure the sustainability of development through environmental protection, create a better ecological environment through sustainable socio-economic development, and realize the coordination, integration and unification of environmental protection and economic development.

Guided by General Secretary Xi Jinping's scientific judgment that the ecological environment is the productivity, we are required to adhere to the concept of green development in practice. Green development serves as the key to accomplishing such arduous tasks as achieving sustainable development of the Chinese nation, social progress, stable economic growth, environmental improvement, and better livelihood. Green development and sustainable development are the trend of the times in the world. The concept of green development in China is highly consistent with the green development and sustainable development advocated by the international community [22]. The green development advocated by the international community is the green transformation of the development model, which is characterized by two major features: relying more on scientific and technological progress. Ecological environment has become a vital component of the comprehensive competitiveness of a country and a region. Green development is the only way to enhance the comprehensive strength and international competitiveness. Therefore, China's green development is the need to ensure that the Chinese nation stands towering among the nations of the word and improve its international competitiveness. It can also display China's positive image to the world, establish a Chinese model of development transformation, provide Chinese experience for the sustainable development of mankind, and make new contributions to global ecological security.

More than that, we need to take immediate action to promote green development with "utmost determination" and explore ways and models of sustainable development and governance. General Secretary Xi Jinping pointed out, "Coordinated and green development are both concepts and measures, and policies must be fully implemented" [23]. The basic of green development is to rationally distribute production, living and ecological space. To pursue green development, we must adhere to the basic state policy of conserving resources and protecting the environment, pursue sustainable development, and foster a new pattern of modernization that promotes harmonious development between man and nature. As a result, the mainstream ways of production and living will transform to resource-saving and environmental-friendly ones. The report to the 19th CPC National Congress pointed out, "We must solve prominent environmental problems. We will get everyone involved in improving the environment and address environmental issues at the root. We will continue our campaign to prevent and control air pollution to make our skies blue again. We will speed up prevention and control of water pollution, and take comprehensive measures to improve river basins and offshore areas. We will strengthen the control of soil pollution and the restoration of polluted soil, intensify the prevention and control of agricultural pollution from non-point sources, and take measures to improve rural living environments. We will improve the treatment of solid waste and garbage. We will enforce stricter pollutants discharge standards and see to it that polluters are held accountable. We will improve our systems for credibility assessment based on environmental protection performance, for mandatory release of environmental information, and for imposing severe punishment for environmental violations" [24], and "implement the strictest possible systems for environmental protection, and develop eco-friendly growth models and ways of life. We must pursue a model of sustainable development featuring increased production, higher living standards, and healthy ecosystems. We must continue the Beautiful China initiative to create good working and living environments for our people and play our part in ensuring global ecological security" [25]. Karl Marx also said, "Thus society is the accomplished union of man with nature, the veritable resurrection of nature" [26]. The relationship between man and nature is not to be master-servant or adversary, but to share the same outlook and destiny.

General Secretary Xi Jinping puts forward "the protection and improvement of ecological environment is the protection and improvement of productive forces". The philosophy points out that ecological progress and the development of productive forces can benefit each other by achieving mutual promotion and coordinated development. On the one hand, the development of productive forces is inseparable from the external ecological environment, which is a decisive factor affecting the structure, layout and scale of productive forces. On the other hand, by utilizing excellent ecological environment, we can develop biological resources development, eco-tourism, environmental protection and other industries according to local conditions on the basis of protection, so as to turn the ecological advantage into economic advantage. In particular, the development of modern economy and society relies more and more on better environment. The better the environment is, the stronger the factors of production can be attracted and condensed into, thus the carrying capacity of economic and social development will be stronger.

General Secretary Xi Jinping's philosophy of "The environment is crucial to the wellbeing of the public and productivity" is a reflection of governing for the people. It inherits and develops Marxism's view of productivity and the concept of man and nature, and makes a more in-depth analysis of the relationship between man and nature. Following General Secretary Xi Jinping's scientific judgement, we must be fully aware of the urgency and difficulty of protecting the ecological environment and tackling environmental pollution, we must be fully aware of the importance and necessity of strengthening ecological progress. In the light of responsible to the people and to future generations, we should liberate and develop ecological productive forces. We must truly resolve to control environmental pollution and improve the ecological environment, strive to move toward a new era of socialist ecological civilization, and create a good production and living environment for the people. A good ecological environment is a necessary condition for sustainable development and an important manifestation of the people's pursuit of a better life.

Guided by General Secretary Xi Jinping's scientific judgment that the ecological environment is the productivity, we are required to adhere to the concept of green development in practice. Green development serves as the key to accomplishing such arduous tasks as achieving sustainable development of the Chinese nation, social progress, stable economic growth, environmental improvement, and better livelihood. Green development and sustainable development are the trend of the times in the world. The concept of green development in China is highly consistent with the green development and sustainable development advocated by the international community [22]. The green development advocated by the international community is the green transformation of the development model, which is characterized by two major features: relying more on scientific and technological progress. Ecological environment has become a vital component of the comprehensive competitiveness of a country and a region. Green development is the only way to enhance the comprehensive strength and international competitiveness. Therefore, China's green development is the need to ensure that the Chinese nation stands towering among the nations of the word and improve its international competitiveness. It can also display China's positive image to the world, establish a Chinese model of development transformation, provide Chinese experience for the sustainable development of mankind, and make new contributions to global ecological security.

More than that, we need to take immediate action to promote green development with "utmost determination" and explore ways and models of sustainable development and governance. General Secretary Xi Jinping pointed out, "Coordinated and green development are both concepts and measures, and policies must be fully implemented" [23]. The basic of green development is to rationally distribute production, living and ecological space. To pursue green development, we must adhere to the basic state policy of conserving resources and protecting the environment, pursue sustainable development, and foster a new pattern of modernization that promotes harmonious development between man and nature. As a result, the mainstream ways of production and living will transform to resource-saving and environmental-friendly ones. The report to the 19th CPC National Congress pointed out, "We must solve prominent environmental problems. We will get everyone involved in improving the environment and address environmental issues at the root. We will continue our campaign to prevent and control air pollution to make our skies blue again. We will speed up prevention and control of water pollution, and take comprehensive measures to improve river basins and offshore areas. We will strengthen the control of soil pollution and the restoration of polluted soil, intensify the prevention and control of agricultural pollution from non-point sources, and take measures to improve rural living environments. We will improve the treatment of solid waste and garbage. We will enforce stricter pollutants discharge standards and see to it that polluters are held accountable. We will improve our systems for credibility assessment based on environmental protection performance, for mandatory release of environmental information, and for imposing severe punishment for environmental violations" [24], and "implement the strictest possible systems for environmental protection, and develop eco-friendly growth models and ways of life. We must pursue a model of sustainable development featuring increased production, higher living standards, and healthy ecosystems. We must continue the Beautiful China initiative to create good working and living environments for our people and play our part in ensuring global ecological security" [25]. Karl Marx also said, "Thus society is the accomplished union of man with nature, the veritable resurrection of nature" [26]. The relationship between man and nature is not to be master-servant or adversary, but to share the same outlook and destiny.

General Secretary Xi Jinping puts forward "the protection and improvement of ecological environment is the protection and improvement of productive forces". The philosophy points out that ecological progress and the development of productive forces can benefit each other by achieving mutual promotion and coordinated development. On the one hand, the development of productive forces is inseparable from the external ecological environment, which is a decisive factor affecting the structure, layout and scale of productive forces. On the other hand, by utilizing excellent ecological environment, we can develop biological resources development, eco-tourism, environmental protection and other industries according to local conditions on the basis of protection, so as to turn the ecological advantage into economic advantage. In particular, the development of modern economy and society relies more and more on better environment. The better the environment is, the stronger the factors of production can be attracted and condensed into, thus the carrying capacity of economic and social development will be stronger.

General Secretary Xi Jinping's philosophy of "The environment is crucial to the wellbeing of the public and productivity" is a reflection of governing for the people. It inherits and develops Marxism's view of productivity and the concept of man and nature, and makes a more in-depth analysis of the relationship between man and nature. Following General Secretary Xi Jinping's scientific judgement, we must be fully aware of the urgency and difficulty of protecting the ecological environment and tackling environmental pollution, we must be fully aware of the importance and necessity of strengthening ecological progress. In the light of responsible to the people and to future generations, we should liberate and develop ecological productive forces. We must truly resolve to control environmental pollution and improve the ecological environment, strive to move toward a new era of socialist ecological civilization, and create a good production and living environment for the people. A good ecological environment is a necessary condition for sustainable development and an important manifestation of the people's pursuit of a better life.

Guided by General Secretary Xi Jinping's scientific judgment that the ecological environment is the productivity, we are required to adhere to the concept of green development in practice. Green development serves as the key to accomplishing such arduous tasks as achieving sustainable development of the Chinese nation, social progress, stable economic growth, environmental improvement, and better livelihood. Green development and sustainable development are the trend of the times in the world. The concept of green development in China is highly consistent with the green development and sustainable development advocated by the international community. The green development advocated by the international community is the green transformation of the development model, which is characterized by two major features: relying more on scientific and technological progress. Ecological environment has become a vital component of the comprehensive competitiveness of a country and a region. Green development is the only way to enhance the comprehensive strength and international competitiveness. Therefore, China's green development is the need to ensure that the Chinese nation stands towering among the nations of the word and improve its international competitiveness. It can also display China's positive image to the world, establish a Chinese model of development transformation, provide Chinese experience for the sustainable development of mankind, and make new contributions to global ecological security.

More than that, we need to take immediate action to promote green development with "utmost determination" and explore ways and models of sustainable development and governance. General Secretary Xi Jinping pointed out, "Coordinated and green development are both concepts and measures, and policies must be fully implemented." The basic of green development is to rationally distribute production, living and ecological space. To pursue green development, we must adhere to the basic state policy of conserving resources and protecting the environment, pursue sustainable development, and foster a new pattern of modernization that promotes harmonious development between man and nature. As a result, the mainstream ways of production and living will transform to resource-saving and environmental-friendly ones. The report to the 19th CPC National Congress pointed out, "We must solve prominent environmental problems. We will get everyone involved in improving the environment and address environmental issues at the root. We will continue our campaign to prevent and control air pollution to make our skies blue again. We will speed up prevention and control of water pollution, and take comprehensive measures to improve river basins and offshore areas. We will strengthen the control of soil pollution and the restoration of polluted soil, intensify the prevention and control of agricultural pollution from non-point sources, and take measures to improve rural living environments. We will improve the treatment of solid waste and garbage. We will enforce stricter pollutants discharge standards and see to it that polluters are held accountable. We will improve our systems for credibility assessment based on environmental protection performance, for mandatory release of environmental information, and for imposing severe punishment for environmental violations", and "implement the strictest possible systems for environmental protection, and develop eco-friendly growth models and ways of life. We must pursue a model of sustainable development featuring increased production, higher living standards, and healthy ecosystems. We must continue the Beautiful China initiative to create good working and living environments for our people and play our part in ensuring global ecological security". Karl Marx also said, "Thus society is the accomplished union of man with nature, the veritable resurrection of nature" [26]. The relationship between man and nature is not to be master-servant or adversary, but to share the same outlook and destiny.

General Secretary Xi Jinping puts forward "the protection and improvement of ecological environment is the protection and improvement of productive forces". The philosophy points out that ecological progress and the development of productive forces can benefit each other by achieving mutual promotion and coordinated development. On the one hand, the development of productive forces is inseparable from the external ecological environment, which is a decisive factor affecting the structure, layout and scale of productive forces. On the other hand, by utilizing excellent ecological environment, we can develop biological resources development, eco-tourism, environmental protection and other industries according to local conditions on the basis of protection, so as to turn the ecological advantage into economic advantage. In particular, the development of modern economy and society relies more and more on better environment. The better the environment is, the stronger the factors of production can be attracted and condensed into, and thus the carrying capacity of economic and social development will be stronger.

General Secretary Xi Jinping's philosophy of "The environment is crucial to the wellbeing of the public and productivity" is a reflection of governing for the people. It inherits and develops Marxism's view of productivity and the concept of man and nature, and makes a more in-depth analysis of the relationship between man and nature. Following General Secretary Xi Jinping's scientific judgement, we must be fully aware of the urgency and difficulty of protecting the ecological environment and tackling environmental pollution, we must be fully aware of the importance and necessity of strengthening ecological progress. In the light of responsible to the people and to future generations, we should liberate and develop ecological productive forces. We must truly resolve to control environmental pollution and improve the ecological environment, strive to move toward a new era of socialist ecological civilization, and create a good production and living environment for the people. A good ecological environment is a necessary condition for sustainable development and an important manifestation of the people's pursuit of a better life.

Guided by General Secretary Xi Jinping's scientific judgment that the ecological environment is the productivity, we are required to adhere to the concept of green development in practice. To achieve sustainable development, social progress, economic growth, environmental improvement, and better livelihood for the Chinese nation, the key to accomplish these arduous tasks is to pursue green development. Green development and sustainable development are the trend of the times in the world. The concept of green development in China is highly consistent with the green development and sustainable development advocated by the international community [3]. The green development advocated by the international community is the green transformation of the development model, which is characterized by two major features: relying more on scientific and technological progress. Ecological environment has become a vital component of the comprehensive competitiveness of a country and a region. Green development is the only way to enhance the comprehensive strength and international competitiveness. Therefore, China's green development is the need to ensure that the Chinese nation stands towering among the nations of the word and improve its international competitiveness. It can also display China's positive image to the world, establish a Chinese model of development transformation, provide Chinese experience for the sustainable development of mankind, and make new contributions to global ecological security.

More than that, we need to take immediate action to promote green development with "utmost determination" and explore ways and models of sustainable development and governance. General Secretary Xi Jinping pointed out, "Coordinated and green development are both concepts and measures, and policies must be fully implemented." The basic of green development is to rationally distribute production, living and ecological space. To pursue green development, we must adhere to the basic state policy of conserving resources and protecting the environment, pursue sustainable development, and foster a new pattern of modernization that promotes harmonious development between man and nature. As a result, the mainstream ways of production and living will transform to resource-saving and environmental-friendly ones. The report to the 19th CPC National Congress pointed out, "We must solve prominent environmental problems. We will get everyone involved in improving the environment and address environmental issues at the root. We will continue our campaign to prevent and control air pollution to make our skies blue again. We will speed up prevention and control of water pollution, and take comprehensive measures to improve river basins and offshore areas. We will strengthen the control of soil pollution and the restoration of polluted soil, intensify the prevention and control of agricultural pollution from non-point sources, and take measures to improve rural living environments. We will improve the treatment of solid waste and garbage. We will enforce stricter pollutants discharge standards and see to it that polluters are held accountable. We will improve our systems for credibility assessment based on environmental protection performance, for mandatory release of environmental information, and for imposing severe punishment for environmental violations" [2], and "implement the strictest possible systems for environmental protection, and develop eco-friendly growth models and ways of life. We must pursue a model of sustainable development featuring increased production, higher living standards, and healthy ecosystems. We must continue the Beautiful China initiative to create good working and living environments for our people and play our part in ensuring global ecological security" [3]. Karl Marx also said, "Thus society is the accomplished union of man with nature, the veritable resurrection of nature" [4]. The relationship between man and nature is not to be master-servant or adversary, but to share the same outlook and destiny.

General Secretary Xi Jinping's philosophy of "by protecting the environment we are protecting productive forces, and by improving the environment we are developing productive forces" pointed out that, the relationship between ecological progress and the development of productive forces is mutual promotion rather than restraint, and it can achieve mutual promotion and coordinated development. On the one hand, the development of productive forces is inseparable from the external ecological environment, which is a decisive factor affecting the structure, layout and scale of productive forces. On the other hand, by utilizing excellent ecological environment, we can develop biological resources development, eco-tourism, environmental protection and other industries according to local conditions on the basis of protection, so as to turn the ecological advantage into economic advantage. In particular, the development of modern economy and society relies more and more on better environment. The better the environment is, the stronger the factors of production can be attracted and condensed into, and thus the carrying capacity of economic and social development will be stronger.

General Secretary Xi Jinping's philosophy of "The environment is crucial to the wellbeing of the public and productivity" is a reflection of exercising power in the interests of the people. It inherits and develops Marxism's view of productivity and the concept of man and nature, and makes a more in-depth analysis of the relationship between man and nature. Following General Secretary Xi Jinping's scientific judgment, we must be fully aware of the urgency and difficulty of protecting the ecological environment and tackling environmental pollution, we must be fully aware of the importance and necessity of strengthening ecological progress. In the light of responsible to the people and to future generations, we should liberate and develop ecological productive forces. We must truly resolve to control environmental pollution and improve the ecological environment, strive to move toward a new era of socialist ecological civilization, and create a good production and living environment for the people. A good ecological environment is a necessary condition for sustainable development and an important manifestation of the people's pursuit of a better life.

Chapter 4
Harmonious Coexistence Between Humans and Nature Must Be Promoted

The contemporary ecological problems are westerners always believe that everything has its vanquisher. They image themselves as conquerors with one technology to overcome another technical problems. The whole industrial production is like a big machine with a tight string, and it goes on and on with part of the equipment for producing products and part of the equipment for purifying wastes. But the result has been to create bigger problems that are harder to solve than ever before. Holmes Rolston, founder of the American Society for Environmental Ethics, pointed out, "traditional western ethics has not considered the value of things other than human subject. ... It seems the Eastern world has a bright future. There is no boundary between facts and values, or between man and nature. In the west, nature, which serves as tool, has been stripped of its intrinsic value" [1]. The significant statement of General Secretary Xi Jinping's "respect nature, follow its ways, and protect it so as to promote harmonious coexistence between man and nature" is not only the inheritance of ecological wisdom and Oriental wisdom of the Chinese ancestors in contemporary China, but also the spiritual home for the Chinese people to recognize the concept of ecological civilization and feel the harmony between man and nature, which provides the ideological and cultural soil for exploring the essence of ecological progress.

1 Ecological Progress Should Be Equipped with Ecosystem View

Ecosystem is an ecological functional unit formed by continuous energy and material exchange between organisms and their abiotic environment (physical environment). For ecosystem, ecological balance is an important condition for the whole biosphere to maintain a normal life support system.

Formed in early days, the ecosystem view is an important part of Xi Jinping's thought on ecological civilization. During Xi's term of office as the secretary of the CPC Ningde Prefectural Committee, Fujian Province in 1989, he integrated the idea

J. Pan, *China's Global Vision for Ecological Civilization*,
https://doi.org/10.1007/978-981-16-4534-1_4

of ecosystem theory in his article *The Basic Skills of Cadres—Maintaining Close Ties with People*, he integrated the idea of ecological system theory in it. He wrote in simple words, "For example, when a dike is built, the problem of unavailable traffic is solved, but the backflow of water is gone, and the ecological balance is destroyed; people are happy to see the problem of disease treatment and bathing is solved by the mass use of geothermal water, however the subsidence of the ground has had even more troubling consequences; so we should not make the same mistake again!".

Xi's view on ecosystem is mainly reflected in three aspects:

The first is to put forward the concept of that "mountains, rivers, forests, fields, lakes and grasslands together form a biotic community". The earth's ecology system includes forests, grasslands, deserts, tundra, swamps, rivers, oceans, lakes, farmland and cities, and many other elements, each of which also constitutes a relatively small sub-ecosystem. General Secretary Xi Jinping believed that in order to protect the ecological balance of a region, we should first consider them as a life community.

The second is to attach importance to the dissemination and practice of the concept of "respecting nature, conforming to nature and protecting nature". The establishment of sustainable ecosystem is an important part of ecological progress. Xi believes that to build a harmonious ecosystem, one of the important prerequisites is to respect nature and not to blindly build an artificial ecosystem. "China will follow the principles of respecting, conforming to and protecting nature, implement the basic state policy of conserving resources and protecting the environment, and more consciously promote green, circular and low-carbon development."

The third is to explore the model of ecological protection and environmental governance from the perspective of system theory. Many aspects are involved in ecological protection and environmental governance. If the policies and measures are not in place, problems will arise as "one solves a problem only to find another cropping up". Xi emphasizes for many times that it is a systematic project for ecological protection and environmental governance.

Xi's thought of ecosystem is a vital part of thought on socialism with Chinese characteristics, of which an overall and perspective innovation is five major vision of "innovative, coordinated, green, open and shared" development. General Secretary Xi Jinping pointed out, "The vision for innovative, coordinated, green, open and shared development is put forward by the 5th plenary session of the 18th CPC Central Committee, which aims to sluggish world economic recovery as China has entered a new normal in economic development. With a firm grasp of the guideline of new development visions, we must firmly practice them. We should immediately adjust our understanding of resolutely correct any behavior and completely abandon any practice that is inappropriate, unsuited or even opposed to the new development visions. In the meantime, inseparable as the five major development visions are, these visions are interconnecting, flowing seamlessly into each other, and mutually reinforcing. These visions are designed to be adhered to and implemented in an integrated, unified fashion. No one of these principles can be emphasized over, or in place of, any of the others.

General Secretary Xi Jinping attaches great importance not only to the elaboration of ecosystem thinking, but also to the application of ecosystem thinking into practice.

From 1973 to 2006, from the Secretary of Village to the Secretary of Provincial Committee of CPC, many practices of Xi have reflected his continuous passion for ecology and his ethical feelings of respecting nature.

- The construction of biogas digester reflects the ecological philosophy. In Liangjiahe village, Shaanxi Province, Xi went to Sichuan Province to learn from their experience at his own expense. He led the villagers to build the first biogas digester in northern Shaanxi, which made the re-utilization of biogas, cooking and lighting accessible. Furthermore, Xi, together with villagers, drilled two wells to turn dry land into irrigated land, followed by the effort to improve soil by constructing five dams altogether aiming for changing geological conditions.
- We prefer less profits to pollution. During Xi's administration of Zhengding County in Hebei Province, in 1985, Xi formulated *General Plan for Economic, Technological and Social Development of Zhengding County*, defining the basic objectives of Zhengding's environmental protection before the end of the twentieth century: to stop the destruction of the natural environment, prevent new pollution and control existing pollution sources. The *Plan* stressed that we prefer less profits to pollution, and pollution sources changing and pollution in villages are strictly prohibited.
- The coordination of social, economic, ecological and other elements should be emphasized to the process of resource utilization. During Xi's administration of Ningde City in Fujian Province, he stressed that resource utilization is not a single process, but is a comprehensive one; it was not only about economic benefits, but also about the coordination of social, economic and ecological benefits. He took the development of forestry resources as a strategic problem for the revitalization of eastern Fujian, citing the words of the masses "we will be prosperous as the mountains in eastern Fujian turn green" to show that the development of forestry is the main way to get rid of poverty and become better off.
- The rise or fall of a civilization is closely tied to its relationship with nature. During Xi's administration in Zhejiang Province, he pointed out that the rise or fall of a civilization is closely tied to its relationship with nature. A government that does not pay attention to ecology is not sober-minded, a leader who does not pay attention to ecology is incompetent, an enterprise that does not pay attention to ecology is hopeless, and a citizen who does not pay attention to ecology is not a citizen with modern civilization consciousness. In 2001, during Xi's term of office as Governor of Fujian Province, he served as the leader himself of the leading group of Fujian Province's ecological progress. He prospectively put forward the strategic conception of building an ecological province and began the largest systematic ecological protection project in the history of Fujian Province.

Since assuming the leadership of China, General Secretary Xi Jinping has given important instructions on the specific work in many aspects in terms of ecosystem. For example, regarding to the development of Yangtze River Economic Belt, Xi pointed out "as basin economy, the Yangtze River Economic Belt involves water, road, port, bank, production, city, biology, wetland, environment and other aspects. It is an entirety which must be comprehensively planned and coordinated. We must enhance our systematic thinking, coordinate the reform and development of all regions, various interregional policies, construction in various fields, and various resource elements. We must make the synergy of provinces and cities along the river, promote the coordinated development of the upper, middle and lower reaches of the Yangtze River Economic Belt, and the interaction and cooperation between the east, middle and western regions, thus making it a pilot demonstration area, innovation-driven area and coordinated development area of China's ecological conservation. We must optimize the efficiency of existing shorelines, integrate water security, flood control, pollution control, port and shore, traffic, landscape, etc., and promptly solve the problem of disordered development of industries and port shorelines along the river. We must optimize the layout of city cluster in Yangtze River Economic Belt, adhere to integration of large, middle and small cities and linkage between eastern, central and western areas, and rely on three major city-clusters: Yangtze River Delta, Middle reaches of Yangtze River and Chengdu-Chongqing to drive its growth."

2 Mountains, Rivers, Forests, Fields, Lakes and Grasslands Together Form a Biotic Community

General Secretary Xi Jinping pointed out, "We must realize that our mountains, rivers, forests, fields and lakes together form a biotic community. Just as humans need the fields for food, the fields need the rivers, the rivers need the mountains, the mountains need the soils, and the soils need the forests. If the mountain is destroyed and the forest is cut down, the water will be destroyed. The mountains will become barren, the water will become floods, causing mud and sand flow together, and the soil will become infertile with soil erosion and ravines." Xi made further statement in the report to the 19th CPC National Congress by adding "grasslands" into "mountains, rivers, forests, fields, and lakes"—"we must implement our fundamental national policy of conserving resources and protecting the environment, and cherish the environment as we cherish our own lives. We will adopt a holistic approach to conserving our mountains, rivers, forests, farmlands, lakes, and grasslands".

In nature, any biological community does not exist in isolation. They are always connected and interacted with their living environment through the exchange of energy and material, forming a unified whole, such a whole is the ecosystem. A complete ecosystem contains elements such as mountains, rivers, forests, fields, lakes and grasslands. Maintaining the balance between these elements is an important guarantee for the sustainable development of human society. Karl Max said, "Nature, as far as it is not the human body itself, is the inorganic body of man. Man lives by nature. That is to say, nature is the human body with which man must interact continuously in order not to die. When the physical and spiritual life of man is connected with nature, it means that nature is connected with itself, for man is a part of nature."

General Secretary Xi Jinping has made several statements and important comments on the importance of various elements such as mountains, rivers, forests, fields, lakes and grasslands to the ecosystem. Firstly, about "mountains". For example, hundreds of private villas were built in the northern Qinling mountain area of Shaanxi Province. The mountains were vandalized. The domestic sewages were discharged at will. In some places, the hillside was even leveled and the forest land was occupied, which caused serious damage to the ecological environment. The civilians have a lot of complains towards the issues. After reading the materials, General Secretary Xi Jinping immediately gave instructions. The illegal buildings, which had existed for years, were demolished in one fell swoop. Secondly is about "water". Forests, lakes and wetlands are natural reservoirs, which have the functions of water conservation, flood storage and waterlogging prevention, water quality and air purification. However, more than 200 lakes covering an area of more than 10 square kilometers have shrunk; nearly 1000 natural lakes have disappeared due to reclamation; and 1.6 trillion cubic meters of precipitation directly flows into the sea every year, which cannot be utilized. In response to this grim situation, Xi pointed out, "if we do not attach importance to protecting the ecological environment such as forests, lakes, wetlands and so on, and then continue to overtake groundwater, the nature will repay greater revenge to us." "One important reason for the scarcity of water is that the ecological space for water conservation have been greatly reduced, and the 'basin' holding water has become smaller and smaller, so precipitation cannot be saved or retained. In the past, we have not done enough systematic research on water regulations. Now, we are focusing on finding new ways to treat the problem from a global perspective, rather than looking for solutions to a problem by appealing to the same ideas." "Tibet need to protect its ecology and China's water tower (the origin of three main rivers in China). We cannot concentrate on trivial matters but neglect the important ones. The ecological problems do more harm than good." Thirdly is about "forests". Xi pointed out, "As the main part of terrestrial ecology, forests, the depending capital of a country and nation, are the foundation of human existence, which is related to the security of survival, fresh water, territory, species and climate, as well as national diplomacy. We must view the issue based on national historical evolution and provide beautiful home for future generations so as to leave positive record for the efforts of contemporary Chinese people." "As the main part and important resource of terrestrial ecology, forests is the significant ecological guarantee for human's survival. It is impossible to imagine what the earth and human beings would be like without forests. For more than 30 years, the implementation of voluntary tree planting has promoted the recovery and development of China's forest resources, and enhanced the people's awareness of cherishing, planting and protecting forests. Meanwhile, we must be aware that China is still a country with few forests and fragile ecology. There is a long way to go for afforestation and ecological improvement." Fourthly is about "fields". Xi pointed out, "Land is the carrier of ecological civilization. In accordance with the principle of balancing population, resources and the environment, and integrating economic, social and ecological benefits, we need to plan the development of China's territorial space as a whole, so as to leave more space for natural restoration." Fifthly is about "lakes". In March 2013, during the

First Session of the 12th National People's Congress, Xi recalled his trip to Suzhou in July 2012 and said, "The beauty of 'heaven' lies in Taihu Lake. Isn't there a song called Beautiful Taihu Lake? It's true that ecology is important. I hope Suzhou can contribute more for the beauty of Taihu Lake." Sixth is about "grasslands". Xi also talks about the issue many times. For example, during his visit to the Inner Mongolia Autonomous Region in 2014, in regard to its ecological environments, Xi pointed out "there are mainly two ways to deal with the ecological environment problems, one is to continue to the implementation of major ecological restoration projects, the control over the sources of sandstorms affecting Beijing and Tianjin, the building to three North Shelter Forest Program and key project for turning marginal farmland into forest or grassland; the other is to actively explore ways to improving system for ensuring ecological progress."

There exists a complex relationship between mountains, rivers, forests, fields, lakes and grassland, which are interdependent and stimulate each other's vitality. They form a community of life organically, and reach a relatively stable state of equilibrium through interaction. If one of the components changes dramatically, it will cause a series of chain reactions, so that the ecological balance is destroyed. Friedrich Engels pointed out, "The whole nature that we come into contact with constitutes a system, that is, the totality of various objects, and the objects we understand here refer to the existence of all materials, from stars to atoms, even to Ether particles, if we admit the existence of Ether particles. These objects are somehow connected, which implies that they interact with each other… It's easy to draw this conclusion as long as we realize that the universe is a system and a totality connected by various objects."

Since the reform and opening up, China has made enormous achievements in short time, which takes hundreds of years for western countries to accomplish. Although the rising momentum is strong, China are at the bottleneck of dealing with development achievements to some extent. We cannot deal with continuous problems through only response to the current situation. The governance ability needs to be integrated. For instance, apart from environmental protection departments, the functions of pollution control are dispersed in department of ocean, harbor supervision, fishery, etc.; the functions of resource protection are dispersed in departments of mineral, forestry, agriculture, water resources, etc.; the functions of comprehensive control and management are dispersed in National Development and Reform Commission, Ministry of Finance, Ministry of Land and Resources, etc. It can be seen from this that much energy need to be exerted in implementing an environmental protection plan which involves several departments. General Secretary Xi Jinping pointed out, "It is very necessary for a department to be responsible for the national land space of utilization control within its territory and the unified protection and restoration of mountains, rivers, forests, fields and lakes." That is to say, although mountains, rivers, forests, fields, lakes and grasslands have their own rights and interests, they are community of life. In the process of development and utilization, we must break through the "game thinking", shelve "departmental interests", form a higher-level coordination mechanism, and integrate all kinds of ecological resources into the framework of unified governance.

3 Respect Nature, Follow Nature's Way and Protect Nature

General Secretary Xi Jinping pointed out, "We should be committed to the philosophy of respecting nature, following its way and protecting it." "To accelerate ecological process, we must fully implement the guiding principles from 18th National Congress. Having been guided by Deng Xiaoping Theory, the Theory of Three Represents, and the Scientific Outlook on Development, we must realize the ecological philosophy of respecting nature, following its way and protecting it. We must implement our fundamental national policy of conserving resources and protecting the environment, act on the principles of prioritizing resource conservation and environmental protection and letting nature restore itself. Efforts should be made to increase awareness of ecological philosophies, improve ecological systems, maintain ecological security and optimize ecological environment, so as to develop spatial layouts, industrial structures, and ways of work and life that help conserve resources and protect the environment."

In the development of human society, the relationship between human and nature is always the eternal theme of human beings. Historically, the evolution of the concept of the relationship between man and nature can be roughly divided into three stages: "Nature decides to win man", "Man decides to win nature" and "Harmony between man and nature". General Secretary Xi Jinping has constantly promoted the harmonious development of man and nature. With his broad mind and global vision, he has realized triple leap from the past, and constantly realizing the new realm of "respecting nature, following its way and protecting it".

The first is the stage of "Nature decides to win man". This mainly exists in the era of relatively backward productivity—mainly refers to the primitive society and agricultural society before the industrial revolution. In this long period, compared with the natural forces, the human forces were vulnerable. Human were neither able to have scientific understanding nor able to gain power to deal with many natural phenomena, and their ability to transform nature is still primary. As Marx said, "At first, nature were opposed to human beings as a completely alien, infinitely powerful and unconquerable force. Human were as much in relation to it as animals were and human obeyed its power as livestock did. Therefore, it is a pure animal consciousness of nature." Due to the incapability of human beings in the face of nature, many phenomena and objects in nature were deified or mysterious, and people appeared relatively weak in front of various "natural gods" (such as Sun God, Moon God, etc.). During this period, the large-scale transformation of nature by human beings was often endowed with mythological features, such as the ancient Chinese mythological stories of King Da Yu taming the flood and old man Yu Gong removing mountains.

The second is the stage of "Man decides to win nature". As we stepped into the era of industrial civilization in 1880s, the Industrial Revolution in Great Britain, marked by the use of spinning Jenny and James Watt's steam engine, not only created the mode of production of large-scale machinery, but also formed a new era of "man decides to win nature" and "combat all forces of nature". Due to the over exploitation and abnormal utilization of natural resources, not only the normal evolution of natural ecology is disturbed, but also balance and stability of the whole natural ecosystem are destroyed, leading to a global biological crisis. Among them, ozone depletion, greenhouse effect, acid rain, desertification and pollution related to water, soil and air have become the global ecological crisis. As a result of deforestation and over-farming, a quarter of the world's arable land have been seriously degraded while more than a third of the land are facing the crisis of desertification. Various kinds of sewage emission and land pollution have made the problems of water pollution become increasingly prominent. At present, nearly a third of the global population are not accessible to safe water, tens of thousands people died every day due to water pollution, and food poisoning is not uncommon. There are many reasons for the serious imbalance between man and nature, but disrespect for the laws of nature, predatory development and excessive development are important factors.

The third is the stage of "Harmony between man and nature". It is because of a deep understanding of the deficiencies of "man decides to win over nature" that China entered the stage of "harmony between man and nature" in the aspect of cognition towards man and nature, and gradually becomes a mainstream thought now. To build a comprehensive and harmonious relationship between man and nature, we need to realize that man is only a part of nature, not the measure of all things. As General Secretary Xi Jinping has pointed out, "It is better to pay the ecological debt sooner rather than later. Otherwise, we cannot conscientious to our generations. If you protect the environment, the environment will be friendly; if you pollute the environment, the environment will revenge mercilessly. This is the objective law of nature, which does not subject to human will. As for the treatment of environmental pollution, we should not be hesitate to pay off the debt with a lot of money."

As far as the category of relationship is concerned, the concept of "Harmony between man and nature" is different from "Nature decides to win man". This is because the cognition of the concept of ecological civilization between man and nature lies in the fact that man can form the experience, culture and consciousness guiding practice through practice. Friedrich Engels pointed out that our whole dominion over nature consists in our superiority over all other creatures and our ability to understand and apply the laws of nature. Day by day, in fact, we learn to understand the laws of nature more correctly and to recognize the recent and further consequences of our interventions in the natural world's habitual processes. Obviously, the concept of "Harmony between man and nature" is different from "Man decides to win nature". This is because the former declared that man need to respect nature, follow nature's ways, and protect nature in the process of nature reformation, which is a reflection on the concept of "anthropocentrism".

In particular, it should be pointed out that in China, although people have gradually realized the importance and necessity of "harmony between man and nature", in practice, pseudo-ecological civilization and even anti-ecological phenomena still exist in large numbers. We believe that the so-called "pseudo ecological civilization construction" is characterized by violating the laws of nature, exceeding the ecological carrying capacity and environmental capacity. For example, in order to beautify the city and improve the urban forest coverage, some cities invest a lot of money to buy trees or even transplant ancient trees to the city. This kind of greening mode, for the original land of big trees or ancient trees, is equivalent to an ecological disaster, and the trees transplanted to a new place may not form a new ecological environment. The second example is that a lot of manpower, money and material resources are invested to many cities which are short of water so as to build artificial lakes, artificial grasslands by sprinkling irrigation and high-specification lawns with high water consumption. This kind of "lucid waters and lush mountains" that violates the laws of nature is also a serious damage to the ecological environment. It goes against the original intention of respecting and conforming to nature, but is also unsustainable. The third example lies in some places under the banner of ecological progress, which "plant lawn today, change to build garden tomorrow, change to plant trees the day after tomorrow. This kind of phenomenon not only does not produce any value, but also costs a lot, which obviously runs counter to the original intention of ecological progress". General Secretary Xi Jinping was keenly aware this kind of phenomenon and pointed out, "Why are so many of cities lack of water? An important reason is that there are too many cement fields, which occupy the forest land, grassland, lake and wetland that can conserve the water source and cut off the natural water cycle, and when the rain comes, it can only be discharged as sewage, leaving less pumped groundwater. To solve the problem of water shortage in cities, we must comply with nature. For example, when upgrading the urban drainage system, it is necessary to give priority to the limited rainwater retention, give priority to the use of more natural forces for drainage, and build a "sponge city" with natural accumulation and purification." Therefore, "Development must be scientific development that follows economic laws, sustainable development that follows natural laws, and inclusive development that follows social laws".

Since the 18th National Congress of the CPC, General Secretary Xi Jinping has made clear demand for several times that we must respect nature, follow its ways, and protect it, and work for a new pattern of harmonious development between man and nature. Take domestic inspection for example, during inspection in Yunnan province, Xi pointed out that rural development should conform to the reality and law of development in rural areas; the rural flavor and landscape which contains homesickness should be retained, and the natural environment and local culture should be protected. Economic development should not come at the cost of destroying the ecological environment. During inspection in Qinghai province, Xi stressed that "we must respect nature, follow its way and protect it, and resolutely work for national ecological security barrier".

The concept of ecological progress is closely related to the earth. On many major international occasions, General Secretary Xi Jinping has also repeatedly spread the concept of harmony between man and nature to convey the firm confidence of China in achieving harmonious development between man and nature. For example, during the General Debate of the 70th Session of the United Nations General Assembly at the UN Headquarters, General Secretary Xi Jinping elaborated on the ecological concept of China's respect for nature and green development, which is to explain to the world China's development goals in the post-industrial era—"we should reconcile industrial development with nature and pursue harmony between man and nature to achieve sustainable development of the world and the all-round development of man", which also express to the world China's determination to shoulder its international responsibilities—"to build a sound ecology is vital for mankind's future. All members of the international community should work together to build a sound global eco-environment. We should firmly respect nature, follow nature's ways and protect nature. We should firmly pursue green, low-carbon, circular and sustainable development. China will shoulder its share of responsibility and continue to play its part in the common endeavor". In November, 2015, Xi gave a keynote speech titled *Work Together to Build a Win–Win, Equitable and Balanced Governance Mechanism on Climate Change* at the opening ceremony of the Paris Conference on Climate Change, which said "All things live in harmony and grow with nourishments. Chinese culture values seeking harmony with nature and respects nature. Going forward, ecological endeavors will feature prominently in China's 13th Five-Year Plan. China will work hard to implement the vision of innovative, coordinated, green, open and inclusive development. China will, on the basis of technological and institutional innovation, adopt new policy measures to improve industrial mix, build low-carbon energy system, develop green building and low-carbon transportation, and build a nation-wide carbon emission trading market so as to foster a new pattern of modernization featuring harmony between man and nature."

4 Systematic Project for Ecological Protection and Environmental Governance

With the enhancement of the ability to transform nature, the activities of human society are exceeding the ability of nature to repair itself, and economic growth and ecological environment protection have gradually become a pair of contradictions. Most developed countries have experienced the development pattern of "Treatment after Pollution". While rich in social wealth, developed countries have also paid a heavy price for ecological environment.

We should learn from what developed countries have been through. However, since the reform and opening up, the problem of environmental pollution in China has become increasingly prominent with economy development. In many places and in many fields, the relationship between economic development and ecological and environmental protection has not been properly handled, and economic development has been achieved at the cost of excessive consumption of resources and destruction of the environment. As a result, energy, resources and ecological and environmental problems have become more and more prominent. Our hope to avoid the pattern of "treatment after pollution" has not been realized, some even become "treatment and pollution". If we do not change this economic development model, it will be hard for resources and environment to support China's sustainable development.

Ecological protection and environmental governance are important part of ecological progress. On many important occasions, General Secretary Xi Jinping has stressed the importance of natural ecological protection and environmental governance.

During his visit in Guangdong Province, Xi earnestly warned, "we owe too much in terms of ecological environment. If we don't hold on to protect environment from now on, we will pay an even steeper price in the future." During his visit to Erhai Lake in Yunnan Province, Xi pointed out that environmental protection is a long-term task and we should be persistent in protecting Erhai Lake so as to let the beautiful scenery forever to be stationed. After taking photos with local officers, Xi said "Sign a note for investigation, I hope the water will be cleaner after a few years."

The long-term practice of local work has made General Secretary Xi Jinping deeply realize that ecological protection and environmental governance is a complex and systematic project. "As systematic project, environmental governance must be treated as a major matter of people's livelihood. We must attach the principle of equal importance to addressing both symptoms and root causes and addressing specific issues, coordination normal governance and emergency emission reduction, mutual promotion of local pollution control and regional development. We must take multiple measures in different places and act together with the whole society." During his inspection in Beijing, Xi pointed out "in a metropolis like Beijing, environmental governance is a systematic project, which must be treated as a major matter of people's livelihood."

General Secretary Xi Jinping's thoughts on systematic ecological protection and environmental governance are mainly reflected in the following aspects:

First, we must attach importance to ecological protection and environmental governance from a political perspective. Xi pointed out that "as the economy goes up, people's sense of happiness has been greatly reduced, and even their sense of discount has been raised up, what's wrong with it? Therefore, we cannot regard the enhancement of ecological progress and ecological environmental protection, and the promotion of a green and low-carbon lifestyle as merely economic issues, but as political issues." "In recent years, Beijing has suffered severe smog, which has seriously affected the health of the people and the image of the Party and China's government."

Second, we must attach importance to strengthen policies, regulations and institutions. Xi believed that "only the strictest system and rule of law can provide a reliable guarantee for the construction of ecological progress. We should establish system of accountability. We must hold those who blindly make decisions regardless of the ecological environment and cause serious consequences accountable for lifelong." "To promote ecological progress and solve the problems of increasingly tight resource, serious environmental pollution and ecosystem degradation, we must take some hard measures and do results-oriented works." The 5th plenary session of the 18th CPC Central Committee pointed out that "we will intensify efforts to improve the environment, with improving the quality of the environment at its core, implement the strictest environmental protection system, fully implement the action plan for preventing and controlling air, water and soil pollution, and implement system whereby environmental protection agencies below the provincial level come directly under the supervision of their respective provincial-level agencies in terms of environmental inspection and monitoring."

Third, we must pay attention to rectify the "GDP-first theory" job performance understanding which only pursues the economic growth rate. For a long time, our ecological environment has been trapped in a vicious circle of "destruction—governance—re-destruction—re-governance". The main reason is that "GDP-first theory" plays a role and we must reshape the understanding of job performance. While attending the specific Party branch meeting of heightening awareness of and implement the Party's mass line for Hebei Provincial and Standing Committee, Xi said that "we should get rid of the stereotype of pursuing GDP. If we have made progress in green development and contributed to the control of air pollution and the solution of smog, we can be proud of ourselves even if the GDP has slipped to the seventh and eighth place. On the other hand, if it is simply for the sake of GDP, but the problem of ecological environment is becoming more and more serious, or the situation is still the same, even if the GDP is booming, we should not agree with it." During his visit in Heixiazi Island, Xi pointed out that "there is no need for the island to be development zone, engineering zone and amusement park. The infrastructure built on the island should be ecologically sound. We should protect the ecology and keep the island clean."

Forth, we must attach importance to the combination of ecological protection, environmental governance to display economic and social benefits. Xi believed that "ecological environment quality is the key to a well-off society. We need to innovate our development ideas and give play to our advantages as a latecomer. We must be responsive to local conditions for developing, give full play to the economic and social benefits of natural environment and ensure that economic, social and ecological benefits are improved simultaneously, so as to achieve the organic unity of people's wealth and ecological progress."

Fifth, we must deliver more to the people that meet their needs. Xi believes that people's demand for ecological products such as fresh air, clear water and environment is more and more urgent, and the ecological environment is more and more precious. We must respond to people's expectations for a sound ecological environment, promote a new approach to green, low-carbon and circular development, and create new growth points from it. The ecological environment is an important task that benefits the country, the people and the future generations. Speaking without practicing is not allowed. Xi cited the simple method for testing environment from ordinary people, "the current standard for netizens to test the water quality of lakes is that mayors dare to jump down and swim or not."

Sixth, we must emphasize the importance and necessity of ecological protection and environmental governance. Xi pointed out that "we should have a clear understanding of the urgency and arduousness of protecting the ecological environment and harnessing environmental pollution, and realize the importance and necessity of ecological progress. With a highly responsible attitude and responsibility toward the people and to future generations, we should make up our mind to control environmental pollution to improve our environment, so as to strive for a new era of ecological progress and create a sound working and living condition for the people."

Seventh, we must take the long term view. Who doesn't study for overall situation is hardly to manage for one aspect. Who has no long-term consideration is hardly to organize temporarily. Xi pointed out that "we must make integrated and sustainable planning on environmental protection. Seeking short-term successes and quick profits are not allowed. The protection of ecological environment is a long-term task, which needed to be done in a long time".

Eighth, we must realize that time waits for no man. "We owe too much in terms of ecological environment. If we don't hold on to protect environment from now on, we will pay an even steeper price in the future."

Ninth, we must pay attention to promoting specific ecological environment engineering. Xi pointed out that "we must implement major ecological restoration projects and enhance the production capacity of ecological products. We must focus on solving prominent environmental problems that do harm to the people, adhere to intensifying the prevention and control and make efforts to control water, air and soil pollution. We will work hard to prevent and control water pollution in key river basins and regions, and work hard to control air pollution in key industries and regions."

Tenth, we must pay attention to comprehensive management. During his visit in Heilongjiang Province, Jilin Province and Liaoning Province in northeast China, Xi stressed that "we should be committed to the strategies for reinvigoration of the old industrial bases in northeast China, and meanwhile the reinvigoration should move forward from a new starting point in new era. We must promote ecological progress, advance comprehensive management measures, fulfill the accountability, advance clean production, expand greenery and make the sky bluer, the mountains greener, the water cleaner and the ecological environment better."

Eleventh, we must pay attention to the specific methods of ecological protection and environmental governance from the micro level. Xi pointed out that "we will step up efforts to control air pollution. The top priority of tackling smog pollution and improving air quality is to control PM2.5. Major measures should be taken to reduce coal consumption, control vehicles' emission, adjust industries' structure, strengthen management, adhere to joint prevention and control, and govern according to law. We need to focus on key areas and key performance indicator, strengthen environmental law enforcement and supervision, and increase efforts to identify accountability."

Twelfth, we must emphasize regional cooperation. Xi pointed out "efforts should be made to expand environmental capacity and ecological space, strengthen cooperation in ecological environment protection, and improve cooperation mechanisms in the fields of shelter forest construction, water resource protection, water environment treatment, clean energy use, etc. on the basis of the cooperative mechanism of air pollution prevention and control that has been launched." Xi also pointed out "we will make greater progress in promoting the development of the Yangtze River Economic Belt. We should take improving the ecological environment of the Yangtze River Basin as the most urgent and important task, and strengthen the restoration of the Basin's ecosystem and the comprehensive management of the environment." at the Forum of the Yangtze Economic Belt initiative in Chongqing.

5 Inheriting Chinese Traditional Ecological Wisdom

It is necessary for us to absorb Chinese excellent traditional culture from the traditional ecological wisdom so as to fully and thoroughly understand and grasp General Secretary Xi Jinping's view of ecological system.

Nowadays, resource constraints are getting tighter, environmental pollutions are getting more and more serious, and ecosystem degradation is in grave difficulties. These problems have attracted much attention from all walks of life. Chinese traditional culture is full of rich ecological wisdom, which still provides profound inspiration and valuable experience for the contemporary ecological progress. General Secretary Xi Jinping has stressed on many occasions that contemporary mankind faces many prominent problems. To solve these problems, we need to use not only the wisdom and strength that mankind has discovered and developed today, but also the wisdom and strength that mankind has accumulated and stored in the history. The excellent Chinese traditional culture, including Confucianism, contains the important enlightenment to solve the problems faced by contemporary human beings. In the face of the prominent problem of the contemporary ecological progress of "the increasingly tense relationship between man and nature", the abundant thoughts of philosophy, humanity, enlightenment and moral concepts of Chinese culture, including "follow nature's course" "man must seek harmony with nature", can provide beneficial inspiration for people to understand and transform the world, can provide useful enlightenment for governance and ethical progress.

One of the basic spirits of Chinese traditional culture is the view on natural peace and harmony, including "a trinity of heaven, earth, and mankind" and "four virtues of great and originating, penetrating, advantageous, correct and firm". The former is the *I Ching's* view on the structure of the universe and the universe as a whole. *The Orderly Sequence of the Hexagrams of I Ching* says, "The three forces are heaven, earth and man"; "Four virtues of great and originating, penetrating, advantageous, correct and firm" are the basic idea of *I Ching* on how to keep all things always harmonious and stable. *The Qian Hexagrams of I Ching* says, "Qian equals to beginning, growth, development and maturity of all things." These two aspects are a harmonious and unified whole. "Heaven and earth existing, all (material) things then got their existence. All (material) things having existence, afterwards there came male and female." Heaven, earth and man are independent of each other as well as "great harmony are preserved in union, the result of which is advantageous, correct and firm." The thought of harmony has nourished the Chinese people with an almost religious sense of compassion for others, so that the Chinese people's cultural genes with the cherishing of natural life, which has had a broad and long-lasting influence on the development of Chinese culture.

"Man must seek harmony with nature" "A trinity with heaven and earth" are most basic Confucian ideas about the relationship between man and nature. Dong Zhongshu, a Han Confucian, said, "When Heaven and man are in harmony, and they become one community of life". Ji Xianlin, a famous linguist, explained it as "Heaven is nature. Man is human beings; Harmony means mutual understanding and friendship between man and nature." But how to achieve harmony between man and nature? The answer is "a trinity with heaven and earth". *The Doctrine of Mean* says, "Only the perfectly sincere person can actualize his own essence. Actualizing his own essence, he can fully actualize the essence of others. Fully actualizing the essence of others, he can fully actualize the essence of all things. Being able to fully actualize the essence of all things, he can assist Heaven and Earth in their transformation and sustenance. Able to assist in Heaven and Earth's transformation and sustenance, he forms a trinity with Heaven and Earth." In today's words, pursuing sustainable natural, economic and social development, which is the supreme goal of human social development. The idea of seeking harmony between man and nature has been an important proposition of Confucianism for 2000 years, establishing the mainstream spirit of Chinese philosophy and tradition, and demonstrating the unique Chinese view of the universe, the unique Chinese values and the unique way of thinking and dealing with problems.

"The law of the Tao is its being what it is" and "The Tao in its regular course does nothing" contain the idea of modern ecological conservation. Chapter 25 of *Tao Te Ching* says, "Man takes his law from the Earth; the Earth takes its law from Heaven; Heaven takes its law from the Tao. The law of the Tao is its being what it is"; Chapter 37 of *Tao Te Ching* says, "The Tao in its regular course does nothing (for the sake of doing it), and so there is nothing which it does not do". The former is the essence of Lao Tzu's thought, which sees the law of nature as the supreme law of all things in the universe and the human world. Lao Tzu believed that the laws of nature must not be violated and that human beings must conform to the ways of heaven. Man can only take his law form the Heaven and Earth. Man must translate the laws of heaven into human norms and conform to the laws of heaven. It is not only the objective law, but also the "supreme virtue" of human beings. The latter is the theoretical basis of Taoism, which points out that the Tao gives birth to all things in heaven and earth and allows them to grow naturally, and its regular course does nothing. However, from the point of view of its results, there is nothing that is not in the order of vitality, and thus the Tao's regular course does nothing but its results does everything. It tells us not to act recklessly, act in force and act in disorder. We should let nature take its course and deal with the relationship between man and nature according to the situation. The 18th National Congress of the CPC calls for the establishment of an ecological conservation that respects nature, follows nature's way and protects nature, which is the inheritance, development and application of the Taoist philosophy of nature.

The Buddhist ideology of "equality of all living things" and "great mercy and compassion" is an important way to achieve ecological progress. Buddhism advocates the equality of all beings. *Treatise on the Mystery of the Mahayana* says, "Not only do sentient beings have Buddha nature, but also grass and trees have Buddha nature. … If all beings become Buddha, all plants and trees will also become Buddha." Buddhism advocates respect for life and against indiscriminate killing and destruction of the ecological balance. Buddhism advocates great mercy and compassion. *Saddharma-pundarika Sutra* (or *Lotus Sutra*) says, "With his great mercy and compassion, he incessantly and indefatigably seeks the welfare of all beings and benefits them all." It is necessary both to quit killing and to protect living creatures. The spirit of "freeing lives" has great practical significance for maintaining ecological balance and preserving the richness and diversity of life. *Convention on Biological Diversity* states at the outset, "State Parties are aware of the intrinsic value of biodiversity ...and they are also aware that the conservation of biodiversity is a common concern of all humanity." In the 1970s, the British philosopher Arnold Toynbee argued, "In the future the world will be unified...I believe it is culture, the unification of culture, that is, Mahayana Buddhism and Confucianism." Although Buddhism is a foreign culture, it has achieved a good integration with the local Chinese culture and has had a great impact on Chinese culture, leaving a splendid Buddhist cultural heritage in Chinese history and becoming an important part of Chinese traditional culture.

Based on the explanations, the unique ecological wisdom or cultural genes of the Chinese nation can be summarized as seeking harmony between man and nature, following nature's course and equality of all living things. The two most remarkable features of this wisdom and genes are, first, that every living individual is able to conform to the natural law through self-moral cultivation and practice so as to achieve "harmonious blend of Heaven and Earth" and "attributes with Heaven and Earth"; second, that he human community should live in harmony with the natural world, which means that heaven is the ultimate source and final destination of human life. Man must obey the natural law, conform to natural law, imitate to natural law, serve to natural law and ultimately achieve the trinity to natural law. A nation with strong cultural creativity is more likely to win the conceptual respect, emotional closeness, and operational support of other nations. Throughout General Secretary Xi Jinping's important remarks and ideas on the concept of ecological conservation, we find that the starting point of the ecological culture to be formed is the requirement in the whole society to establish a new cognitive outlook on ecological civilization that "respects, complies with and protects nature". And in the new era, the sublimation of traditional Chinese civilization has been achieved with new civilizational ideas and cultural forms, thus laying the foundation for the comprehensive rejuvenation of the Chinese nation in terms of ecological culture and green civilization. At the same time, it has also "contributed Chinese wisdom and Chinese solutions to solving human problems".

Chapter 5
Ensuring the Needs for Ecology and Development

General Secretary Xi Jinping pointed out "we must think about the worst-case scenario from the perspective of bottom-line thinking, strive for the best results and be more prepared and advantageous against trouble." The bottom-line thinking is a kind of positive thinking ability that objectively sets the lowest goal, stands at the lowest point and strives for the maximum expectation. Comrade Mao Zedong said, "no matter what work is, we should think about it from the worst possibilities and make plan for it." At present, although China has made some achievements in ecological progress, it is more important for us to pay more attentions to the problems and realize that the overall trend of ecological environment deterioration in China has not been fundamentally reversed. In the face of many problems, we must step up our efforts to resolutely curb the trend of ecological environment deterioration and gradually improve and optimize the ecological environment. But at the same time, we must be soberly aware that China is still and will be in the primary stage of socialism for a long time. Development is the top priority and remains the key to solving all our problems. Xi said we must ensure the needs for ecology and environment. This means we must attach equal importance to ecology and development. We must deal with these two issues at once and attach sufficient importance to both.

1 Two Bottom Lines of Ecological Progress

Ensuring the needs for ecology and development is an important part of Xi Jinping's thought on ecological progress. It is commonly referred to as "the two bottom lines" and is the specific content of Xi's bottom line thinking. "Bottom lines for ecology" refers to the maximum level of our environment and natural resources can bear. "Bottom lines for development" refers to the economic development level needed for poverty alleviation in contemporary China. We cannot realize ecological progress without development.

J. Pan, *China's Global Vision for Ecological Civilization*,
https://doi.org/10.1007/978-981-16-4534-1_5

General Secretary Xi Jinping's thought on ecological progress attaches both importance to the protection of ecological environment and development. The concept of Xi's "two bottom lines" was originally proposed to Guizhou Province. In November 2013, when Xi listened to the local report, he pointed out that Guizhou should ensuring the needs for development and ecological protection. Later, he made more clear requirements towards Guizhou, "we must correctly handle the relationship between development and ecological environment protection, take the lead in the reform of the system and mechanism of ecological civilization construction, implement the proposed action plan into actions, and realize the coordinated promotion of development and ecological and environmental protection." "The proper handling of the relationship between environmental protection and development equals to the relationship between mountains of green and mountains of gold, which is the inherent requirement for sustainable development and major issues of principle for promoting modernization."

To ensure the needs for development and ecology, this is not only General Secretary Xi Jinping's clear request for the economic and social development of Guizhou, but also the ardent hope for all parts of the country. Among Xi's two bottom line concepts, the more important ones are mainly reflected in three areas:

First is that development can be accelerated and the ecology can be protected at the same time. There is a certain contradiction between development and ecological protection, but there is a certain compatibility between the two, and the key lies in the choice of development model and development path. According to General Secretary Xi Jinping, "As long as the guiding principle is right and the relationship between the two is well grasped and handled, it is possible to both accelerate development and safeguard the ecology."

Second is that the concept of red lines for ecology should be firmly kept in our mind. Red lines for ecology is the bottom line and lifeline of national ecological security. General Secretary Xi Jinping stressed, "The concept of the ecological red line must be firmly established. Our ecological problems have reached such a serious level that the most severe measures must be taken, otherwise not only will it be difficult to fundamentally reverse the general trend of ecological degradation, but the other ecological development goals we envisage will also be difficult to achieve. It is important to carefully study and argue what should be included in the ecological red line and how to protect it institutionally and protect the sound ecosystem as much as possible. When it comes to ecological environmental protection, it's about not going one step beyond the minefield or you should be punished."

Third is that bottom line for ecology equals to bottom line for development. Development, is to safeguard and enhance the quality of life. And the maintenance and improvement of bottom line for ecology is also the content, goal and baseline of development. Ensuring that the people can breathe fresh air, drink clean water and eat food with peace of mind is essentially the bottom line of environmental quality safety, the bottom line of safeguarding the basic environmental quality needs for human survival, which is also a theme that has always been of great concern to General Secretary Xi Jinping.

The concept of "two bottom lines" is an important part of Xi's "bottom line" thinking. "Bottom line" thinking is a distinctive theoretical feature of Xi Jinping's socialist ideology of Chinese characteristics in the new era, and since the 18th National Congress of the CPC, General Secretary Xi Jinping has repeatedly emphasized bottom-line thinking and the need to accomplish duties. We must think about the worst-case scenario from the perspective of bottom-line thinking, strive for the best results and be more prepared and advantageous against trouble.

As far as the "bottom line" itself is concerned, there are three main understandings: first, the bottom line is understood as a relatively vague objective claim, with certain limits; second, the bottom line is understood as a numerical concept, referring primarily to the threshold at which things are within capacity; and third, the bottom line is understood as a spatial concept, referring primarily to the spatial boundaries that are drawn. The bottom-line thinking of Xi Jinping's socialist thinking with Chinese characteristics in the new era is mainly reflected in the following aspects:

First is the legal bottom line. The law embodies the will of the State and is the bottom line of the norm of conduct for all citizens. The law is the unified expression of the Party's proposition and the will of the people, and everyone has to operate within the law. General Secretary Xi Jinping pointed out: "Leading cadres should remember that the red line of the law cannot be exceeded, the bottom line of the law cannot be crossed. We must take the lead in observing and enforcing the law and in creating a legal environment in which we work in accordance with the law and resolve problems and conflicts through law. Planning should be done with a rule of law mindset, dealing with problems should be done with a rule of law approach, and considering whether it is legal before we speak or act." "Leading officials at all levels shoulder important responsibilities in advancing the rule of law. We must firmly establish the concept that the red line of the law cannot be crossed and that the bottom line of the law cannot be exceeded; we must not interfere with judicial activities, intervene in the handling of specific cases and interfere improperly in the work of the judiciary authorities." "Those who violate the rules of discipline, violate the laws and regulations, cross the 'red line', cross the 'bottom line', or step into 'minefield', should be investigated firmly and seriously. We should not break the rules because of powerful officials, appease the problem because of minor crime and let the violators off because of numerous numbers of them. Pulling strings is not allowed. We should prevent visible signs of disorder and misbehavior, otherwise they may encourage further disorder and misbehavior, leading to serious crimes."

The second is the bottom line of discipline. A country has its own laws and a family has its own rules. The rules and regulations are the premise and basis for the bottom line of the law to be followed. A department, a unit and a group should formulate the corresponding laws and regulations to restrict and regulate the behavior of internal members according to their own characteristics and reality. As the ruling party, the CPC's rules and disciplines reflect the Party's ideals, beliefs and purposes, and are also the ruler of the Party and the insurmountable bottom line for Party members. *Regulations on CPC disciplinary action* integrates the requirements of the Party Constitution on discipline into political discipline, organizational discipline, integrity discipline, mass discipline, work discipline, life discipline, lists a negative list, focuses on the establishment of rules and regulations, and draws out the untouchable bottom line of Party organizations and members.

The third is the bottom line of policy. In order to realize the interests and aspirations of the classes and strata represented by the organs of State power, political party organizations and other socio-political groups, policies are standardized in the form of authority, which comprise a set of goals to be achieved, principle of actions to be followed, tasks to be accomplished, work patterns to be implemented and general steps and concrete measures to be taken within a certain historical period. The State's policies are the guidelines for national action and the bottom line that cannot be violated or modified by organizations and individuals at all levels. General Secretary Xi Jinping pointed out, "We must respond to what the CPC Central Committee advocated. We must comply with what the CPC Central Committee determined. We must eradicate what the CPC Central Committee banned. Countermeasure action in procession of policy implementation, disregard of orders and prohibitions and compromises in the implementation of central decision-making and deployment are not allowed."

The forth is the moral bottom line. The moral bottom line refers to the minimum threshold of social morality that people should follow, which is the minimum basic norm of morality, i.e. the minimum moral requirements for the subject of the act. For the members of society, the moral bottom line is: honesty, generosity and conscience, without harming others and society, and the law is the minimum moral requirement. The CPC represents the development requirements of China's advanced productive forces, the orientation of China's advanced culture, and the fundamental interests of the Chinese people. For party members and party cadres, they should not only uphold bottom lines of law, discipline and policy, but also uphold moral bottom line for the people. General Secretary Xi Jinping pointed out, "The key to the integrity and self-discipline of cadres lies in holding the bottom line. As long as we can keep the bottom line of conducting oneself, dealing with affairs, exercising authority and making friends, we can uphold the political responsibility entrusted to ourselves by the Party and the people, safeguard our political lifeline, and keep the right values of life. All leading officials must take the fight against corruption and the promotion of integrity as a compulsory course in politics, and must not turn power into a tool for seeking personal or minority interests. We must always preserve the political nature of the Communists." Xi demanded that all the Party and government cadres "should establish correct outlook on world, life and value, uphold Communism, stick to ideals and beliefs, and be attentive to people's feelings for the benefit of the people. We must be self-disciplined, adhere to bottom line, and abide by law. We must be diligent in government works and shoulder the responsibility for promoting reform, development, stability, and continuous improvement of people's lives."

2 Accelerating Development and Protecting Ecology

An accurate understanding of the relationship between the "two bottom lines" is a prerequisite for ensuring the "two bottom lines" of development and ecology. Development is the absolute principle and the top priority of the party in governing and rejuvenating the country. Ensuring the needs for development is also the important guarantee for ensuring the needs for ecology. The 18th National Congress of the CPC emphasized that taking economic construction as the center is the key to rejuvenating the country and development is still the key to solving all problems in China. At present, although our country's total economic output has leapfrogged to the second largest in the world and its influence and pulling power on the world economy has been increasing, we are still a developing country with a large gap between the per capita GDP level and the world average, which is less than one fifth of the per capita level of developed countries, and there is still a clear gap between the overall economic competitiveness and that of developed countries. Therefore, only by promoting sustained and sound economic development can we build a solid material foundation for national prosperity, people's well-being and social harmony and stability. The whole party and the country must adhere to the strategic idea that development is the absolute principle and must not waver in the slightest in this pursuit.

Without development, there can be no real sense of the Chinese dream of a well-off society and a great rejuvenation of the Chinese nation. At the same time, in order to protect the environment, repair the ecology, build a better ecology and realize the blue sky, green earth and clear water, we need a large and continuous ecological input. These ecological inputs can only derive from continuous development. Without sound development as the basis, it is just an empty talk to vigorously promote ecological restoration and ecological development.

At the same time, the development of the ecological environment is an important supporting force for economic development, and economic development without environmental protection is to drain the pond to get all fish. For a long time, under the guidance of the concept of keeping economic development as our central task, many places have weakened their efforts to protect the environment, exchanging economic development for uncontrolled consumption of resources and destruction of the environment, resulting in increasingly prominent problems of energy resources and ecological environment. As a result, the lack of space and momentum for socio-economic development makes sustainable development unsustainable. General Secretary Xi Jinping has sharply criticized the misconception of some cadres that "development is the absolute principle" is interpreted one-sidedly as "economic growth is the absolute principle" and GDP determines everything. In his believe, development, in the final analysis, is for the overall progress of society and the continuous improvement of people's living standards, emphasizing that economic growth is not the same as economic development, that economic development is not simply a matter of speed, that economic development does not represent comprehensive development, and that it should not come at the expense of the ecological environment. One of the most important aspects of putting people first is that the environment in which people themselves live must not be destroyed in the course of development. If there is a serious deviation in population, resources and environment, who can live and work in peace and contentment? And how can a harmonious society be achieved? What's the point of achieving anything else when people are having trouble surviving? In the process of pursuing environmental protection and ecological progress, if we take actions earlier, we will get twice the result with half the effort; if we take actions late, we will get half the result with twice the effort. Development that carves for profits, that pollutes first and then regulates, destroys first and then restores must be banned. "We need to vigorously protect the ecological environment and achieve leapfrog development and coordinated progress of the ecological environment."

At present, the ecological environment problem that our country faces mainly includes: first one is strengthening energy and resources constrains. The problems of a large number of people with relatively little arable land and lacking water resources are becoming more and more acute, making it increasingly difficult to guarantee the security of energy and important mineral resources. Second is relatively serious environmental pollution. A significant portion of our cities cannot meet the new air quality standards. The frequent occurrence of haze weather in the eastern and central regions, especially in Beijing-Tianjin-Hebei region and the surrounding areas, which are larger in size, longer in time and more seriously polluted, have highlighted the serious situation of air pollution in China. The problems of river systems, groundwater pollution and drinking water safety in the country cannot be ignored; in some areas, heavy metals and soil pollution are more serious. Third is the prominent problem of ecosystem degradation. China's forest coverage is low. The area of water loss and soil erosion, desertification land and degraded grasslands are relatively large. Natural wetlands are shrinking, the ecological functions of rivers and lakes are degenerating, and biodiversity is showing a declining trend. Forth is the unreasonable pattern of land development. In general, there are problems such as too much space for production, too little for ecological and living space. Some areas are approaching or exceeding the limits of the carrying capacity of resources and the environment due to blind, over and disordered exploitation. Fifth is the new challenges in addressing climate change. China's total greenhouse gas emissions are large and the task of reducing them is arduous. Sixth is the prominent social impact of environmental problems. Some enterprises have caused environmental pollution by illegal emissions, which causes multi-echo among the public and society.

That is to say, we need to ensure the need for both development and ecology. Development is the fundamental factor in solving many of China's problems; without sustained development, employment and income will be impossible, social stability will be in jeopardy, and there will be no guarantee for various reforms. Therefore, maintaining a certain speed of development is the bottom line of our country's development. At the same time, it is equally important to hold the ecological bottom line, especially now, the carrying capacity of the environment in some places has reached or is close to the critical point, if we put on the last straw, not only the environmental problems are causing trouble, but also will shake the foundation of economic development.

Ensuring two bottom lines requires us to prevent the tendency to do nothing. For example, due to the fear of getting into trouble, one may take bottom line as 'shield' and escape from solving problems, actions were not taken towards those need to be changed, explored and tested. On this issue, General Secretary Xi Jinping said: "It is right to emphasize that development cannot destroy the ecological environment, but it is a bit absolute to dare not take steps towards development in order to protect the ecological environment. In fact, with the right guiding ideology, with the right grasp and handling of the relationship between the two, it is possible to both accelerate development and protect the ecology."

In order to keep the two bottom lines of ecology and development, we must handle the relationship between ecology and development properly, so as to stick together to achieve mutual benefits. Take Guizhou Province's development thinking to safeguard the two bottom-line development goals of ecology and development as an example.

Guizhou Province is determined to firmly adhere to the two bottom lines of development and ecology, not to take the old pattern of pollution first and then governance, not to take the old pattern of sacrificing the environment in exchange for a momentary increase in GDP, nor to live a poor life with clear waters and green mountains. Guizhou is committed to a new pattern of ecological priority, green development, well-off life and ecological beauty. The bottom line is to adhere to the ecological priority and green development. Guizhou is adhere to the foot bottom lines of green mountains, blue sky, clear water and beautiful landscape.

The specific objective is that, given the relatively small size of the economy and its relatively weak development base, maintaining a relatively rapid pace of development for a certain period of time is the bottom line for development, while not increasing backward production capacity and destroying the ecological environment is the bottom line for ecology. To this end, the experience of Guizhou Province is to grasp three points: first, firmly grasp the concept that development is the top priority. Guizhou is adhere to the main theme and the general tone of the work of seeking progress in a steady manner, takes improving the economic quality and efficiency as the center, pays close attention to the economic development wholeheartedly, and makes every effort to create a new period of sustainable development. Second is to vigorously promote ecological progress. In order to ensure the overall stability of environmental quality of the whole province, Guizhou establishes ecological pilot zone, strict ecological protection red line system and implement real-time monitoring of all cities and counties. Third is to take a new pattern of industrialization. Guizhou is leading the industrialization, information integration and transformation development with scientific and technological innovation and independent innovation. According to the actual situation, Guizhou proposes to focus on industry instead of energy and mining industry, promote the ecological development of traditional industries, the large-scale development of characteristic industries and the high-end development of emerging industries, and to promote circular, clean and green development.

3 Idea of Ecological Red Lines with Bottom-Line Thinking

A similar expression to the ecological bottom line is the term "ecological red line". The ecological red line is the bottom line and lifeline of national ecological security, which cannot be breached, and once breached, ecological security, people's productive lives and sustainable national development will be endangered. Our country's ecological problems have reached such a serious level that the most severe measures must be taken, otherwise not only will it be difficult to fundamentally reverse the general trend of ecological degradation, but the other ecological development goals we envisage will also be difficult to achieve.

Ecological red line is one of China's three important control lines. General Secretary Xi Jinping pointed out in the reports of the 19th National Congress of the CPC, "We will complete work on drawing redlines for protecting the ecosystems, designating permanent basic cropland, and delineating boundaries for urban development."

In terms of its basic connotation, the ecological red line is mainly defined in three aspects: first, the spatial red line. The Ministry of Environmental Protection has issued the *Technical Guideline Document for China's Ecological Protection Red Line Delineation* which refers to the strict control and regulation of boundaries in areas such as key ecological function areas, ecologically sensitive areas and vulnerable areas, and is the bottom line of national and regional ecological security. This is because the reproduction of biological populations requires a certain space to have a basic food supply and security of movement. Siberian tiger conservation, for example, which requires hundreds of square kilometers of natural space to live, survive and reproduce. The protection of drinking water sources must also be spatially delineated to protect the water quality from pollution and damage. The red line for arable land is actually a red line for spatial control as well. The second is the red line of resource consumption and environmental quality. A breach of the red line on resource depletion, which caps the use of resources, especially renewable resources, such as water utilization, timber harvesting, fishing capacity, etc., would result in resource decay and system collapse. The quality red line is based on human and ecosystem health needs. The atmospheric PM2.5 concentration level, the concentration of Chemical Oxygen Demand (COD) and the level of heavy metal in soil are all related to human healthy life and must meet certain quality red lines. The third is standards set by policy provisions, such as emission standards, technical standards and other policy red lines, so as to meet and highlight spatial and quality red lines. Spatial red lines and quality red lines are the product of scientific verification and measurement, and are not flexible; however, standard policy red lines are determined according to industry characteristics, level of development, technological capabilities, such as energy consumption per unit of GDP, carbon dioxide emissions per unit of GDP, which play roles as tools and measures. The spatial concept of red lines for ecological protection is aimed at maintaining national or regional ecological security and sustainable development, with areas requiring special protection according to the need to protect ecosystem integrity and connectivity; The quality red line of environmental medium (space, water, soil) is drawn mainly for the health and safety of human beings and ecosystems to prevent the decline or even collapse of ecosystems. Once the ecological red line is breached, the ecological balance will inevitably be disrupted, even with catastrophic consequences, and it will often be difficult to restore the original status quo ante, even if a lot of manpower, financial and material resources are invested.

There is a close link between the "ecological bottom line" and the "ecological red line". First of all, in the context of "ecological red line" as a spatial concept, the demarcation of red line area is an important way to protect a regional ecological bottom line. The whole country should defend the ecological bottom line, therefore, each region should be regarded as the bottom line. However, as many places need to be developed, few areas are demarcated as red line areas that are strictly protected by policy and regulations that restrict development or prohibit development. The development and use of water, land, forest, energy and other resources are strictly protected in the delineated red line area, which is intended to strictly protect the ecological bottom line of the area. Secondly, the "ecological red line" established by the threshold level of environmental media is an important basis for determining the red line. In the field of ecological environment, the ultimate limit of its carrying capacity should be selected as the bottom line. Once the bottom line is breached, the situation will be irreversible and needs to be safeguarded by the use of policies and regulations, with the binding force of policies and regulations, the bottom line also becomes a red line. As the bottom line is flexible, in some areas where ecological protection and resource utilization are better, the bottom line standards for some indicators are relatively high, and the red line is also relatively high. For some areas with serious ecological damage and poor resource utilization, the bottom line and red line of some indicators can also be adopted standard of "improving rather than degenerating" within a period of time. But in either case, the bottom line value is an important basis for determining the red line value. People often say the red line is the bottom line, which is mainly based on this situation.

In essence, the essence of the ecological protection red line is the bottom line of ecological and environmental security, which aims to establish the most stringent ecological protection system and put forward higher regulatory requirements for ecological function protection, environmental quality security and natural resource utilization, thereby promoting a balanced population, resources and environment and the unification of economic and social ecological benefits.

General Secretary Xi Jinping attaches great importance to promoting the whole society to keep ecological red line concept in mind, "We must firmly establish the concept of ecological red line. When it comes to ecological and environmental protection, we must not cross the line, or we will be punished." He also emphasizes the importance of the implementation of the concept and the path, "We must carefully study and argue what should be included in the ecological red line and how to protect the ecological red line from the system." On the development of Shaanxi-Gansu-Ningxia Old Revolutionary Bases, Xi said, "To promote the development of the Shaanxi-Gansu-Ningxia Old Revolutionary Bases, we must combine with natural conditions and resource distribution, make scientific and rational planning.

In the process of development, we must resolutely uphold the ecological red line so that the clear sky, green grassland, and flocks of herds are always the characteristic scenery of the Loess Plateau." On the protection of snow-covered plateau, Xi said, "We must strictly enforce the bottom line, red line and top prohibition of ecological safety, improve the comprehensive ecological compensation mechanism, effectively protect the snowy plateau and build national ecological security barrier." The reason why Xi stressed the need to firmly establish the concept of the ecological red line and not to cross the line, is that the ecological red line is the bottom line and lifeline of national ecological security. This red line cannot be breached. Once broken, it will make a threat to ecological security, people's production and living, and national sustainable development.

The "red line of ecological protection" is another "lifeline" mentioned at the national level, following the "one billion and eight hundred million acres of arable land red line". In November 2013, General Secretary Xi Jinping proposed in the *Decision of the Central Committee of the Communist Party of China on Some Major Issues Concerning Comprehensively Deepening the Reform*, which he presided over, that "We must delimit the red line for ecological protection. We will unwaveringly implement the system of functional areas, establish the system of land and space development and protection, promote development in strict accordance with the functional definition of these functional areas, and establish the national park system. We will establish the mechanism to monitor and give early warning to the carrying capacities of resources and the environment, and implement restrictive measures for regions where water and land resources, the environment and oceanic resources have been excessively exploited. We will cancel regional GDP assessment for regions under restrictive ecological development and for national key counties for poverty alleviation displaying ecological fragility."

In March 2015, a meeting was held by the Political Bureau of the CPC Central Committee and presided over by General Secretary Xi Jinping, the meeting adopted *Opinions on Accelerating Ecological Progress*, which clearly proposed that the ecological red line of resources and environment should be strictly observed, and we should establish bottom-line thinking, set and strictly adhere to the upper limit of resource consumption, the bottom line of environmental quality and the red line of ecological protection, and limit all kinds of development activities to the carrying capacity of resources and environment.

In order to implement relevant demands of resources and environment ecological red line of *Opinions of the CPC Central Committee and the State Council on Accelerating Ecological Progress*, so as to guide delineation of the red line, promote the establishment of red line control system and accelerate ecological progress, in May 2016, the National Development and Reform Commission and other nine ministries issued the *Guidance on Strengthening the Resource and Environmental Ecological Red Line Control*, in which the "resource and environmental ecological red line control" is clearly proposed and defined.

The control covers three aspects:

First is to determine the upper limit of resource consumption, which is, to set a reasonable level of national and regional resource consumption, the implementation of control over the total consumption of energy, water, land and other strategic resource consumption, and strengthen the synergy between the control of total resource consumption and the management of consumption intensity. In particular, we should set an upper limit of resource consumption and formulate an effective management system.

Second is to observe the bottom line of environmental quality, which adhere to the principle of "improving rather than degenerating". In accordance with the principles of putting people first, integrated prevention and treatment, addressing both symptoms and root causes, and taking comprehensive measures, an environmental management system should be established with human health as the core, environmental quality improvement as the goal and environmental risk prevention and control as the baseline. We will take the implementation of the "Action Plan of Air, Water and Soil Pollution Prevention and Control" as opportunities, vigorously combat atmospheric, water and soil pollution, and strive to improve environmental quality and strictly control unexpected environmental risks.

Third is to delimit red line for ecological protection and resolutely curb the trend of ecosystem degradation. At present, it is necessary to determine the scope of the red line of ecological protection in accordance with the law as soon as possible, to demarcate the boundary of the red line of ecological protection reasonably, and to play the role of the red line of ecological protection as the baseline in the multiple regulations.

The realization of red lines for ceilings on resource utilization, minimum standards for environmental quality, and environmental protection requires a sound institutional guarantee. General Secretary Xi Jinping said: "The key to whether ecological environmental protection can be implemented in practice lies in the leading cadres. We must adhere to the principles of adhering to the law and regulations, following objective, fair and scientific actions, integrating power and responsibility, and abiding by lifelong accountability. We must focus on the requirements of strictly observing the upper limit of resource consumption, the bottom line of environmental quality, and the red line of ecological protection, and clarify the accountability of leading officials at all levels in decision-making, implementation, and supervision. Leading cadres responsible for causing ecological and environmental damage must be held accountable, whether they have been transferred, promoted or retired. Party committees and governments at all levels should attach practical importance to these issues and strengthen management. The discipline inspection and supervision organs, organizational departments and relevant regulatory departments of the government should fulfill their duties to form synergy." In accordance with the clear requirements of the Party Central Committee and State Council, the *Measures for Accountability for Party and Government Leading Cadres of Ecological and Environmental Damage (Trail Edition)* sets "we must establish bottom-line thinking, set and strictly abide by red lines for ceilings on resource utilization, minimum standards for environmental quality, and environmental protection", highlight the main contradictory aspects, and set up accountability for the performance of duties by leading party and government cadres with a strong negative impact on the ecological environment and strong social reflection.

In addition to minimum standards for environmental quality, and environmental protection mentioned many times before, General Secretary Xi Jinping has always attached importance to strict adherence to resource consumption limits. He pointed out, "From the perspective of resource and environmental constraints, the space for energy resources and ecological environment were relatively large in the past, but now the carrying capacity of the environment has reached or is close to the upper limit, we must meet the expectations of the people for a sound ecological environment and promote the formation of a new way of green and low-carbon circular development." "Resource conservation is fundamental to the protection of the ecological environment. We will vigorously conserve and intensify the resources utilization, promote a fundamental change in the way resources utilization,, substantially reduce the intensity of energy, water and land consumption, vigorously develop a circular economy, and promote the reduction, reuse and recycling of production, circulation and consumption. We need to stick to the basic state policy of resource conservation and environmental protection, act on the principles of prioritizing resource conservation and environmental protection and letting nature restore itself. Efforts should be made to increase awareness of ecological philosophies, improve ecological systems, maintain ecological security and optimize ecological environment, so as to develop spatial layouts, industrial structures, and ways of work and life that help conserve resources and protect the environment."

4 Solving Prominent Environmental Problems

General Secretary Xi Jinping made clear statement in the report to the 19th National Congress of the CPC, "we must solve prominent environmental problems." Ensuring that people can breathe fresh air, drink clean water and eat safe food is essentially the bottom line of environmental quality and the safety line for maintaining the basic environmental quality needs of human existence, including the red line for meeting environmental quality standards and the red line for controlling total pollutant emissions. The red line of environmental quality requires that all environmental elements should meet the standards of the environmental functional area, that is, the quality of the atmospheric environment, the quality of the water environment, the quality of the soil environment, etc., all of them must meet national standards to ensure the safety and health of the people. The red line for the control of total emissions of pollutants is the requirement for fulfilling the task of emission reduction, effective control and reduction of total emissions of pollutants.

General Secretary Xi Jinping has always been concerned about how to make "fresh air available to the people". As early as 2005, Xi Jinping, the then Secretary of Zhejiang Provincial Committee of the CPC, stressed during a forum on the work of population resources and environment, "We must let the people drink clean water, breathe clean air and eat safe food through the construction of ecological progress in the province." Since the 18th National Congress of the CPC, Xi has paid more attention to this issue. At the 5th plenary session of the 18th CPC Central Committee, Xi pointed out, "The people's demand for clean air, clean drinking water, safe food, and a beautiful environment is growing stronger." He said, "The most basic need of deep breathing has become the most blissful pursuit of the people nowadays, and it is worthy of our deep reflection." While participating in the deliberations of the Guizhou delegation at the 2nd session of the 12th National Congress of the CPC, General Secretary Xi Jinping said that air quality is directly related to the happiness of the general public, adding humorously that "air cans" could be made in Guizhou. As for air governance, Xi emphasized that "We cannot take the advantage of other driving forces as success comes with dedicated efforts."

"How to make clean water available to the people" has also been a long-term concern of General Secretary Xi Jinping. Since the 18th National Congress of the CPC, Xi has repeatedly made important remarks on security of drinking water, given that there are still hundreds of millions of rural residents in the country with drinking water security problems. He stressed the need to ensure the safety of drinking water for urban and rural residents, especially in resettlement and rural areas, so that the people can drink clean, safe, assured water. "Drinking water insecurity cannot be brought into a well-off society." In 2014, the media reported herdsman in the hinterland of Tengri desert in the Inner Mongolia autonomous region saying that local enterprises discharged untreated wastewater into drying ponds, and that the wastewater sludge from the drying ponds was directly shoveled out and buried in the desert. In response, General Secretary Xi Jinping made important instructions in September, October and December 2014, and in December 2014, after Xi's instructions, the State Council set up a special inspection team to urge the Tengri Industrial Park to carry out large-scale rectification. A total of 24 persons responsible have been held accountable in the Inner Mongolia autonomous region.

Emphasizing drinking water security is an important part of Xi's thinking on water security. Xi has always paid attention to water security and constantly pointed out, "Crude oil can be imported, there are alternative energy sources on top when the world's oil resources run out, but without water, where can we import it?" "Water security is an important matter involving the long-term security of the country, the whole party should vigorously enhance the awareness of water concerning and water crisis, and attach great importance to solving the problem of water security from the strategic height of realizing a well-off society and achieving the sustainable development of the Chinese nation." On the issue of water safety and security, Xi has proposed such development ideas as "water to fix the city, land, people and production" and "giving priority to water conservation, adhering to spatial balance, systematic management and two-pronged efforts".

"Eat at ease" and "food safety" have also been the focus of General Secretary Xi Jinping. In 2013, during his visit to Qing Feng Steamed Stuffed Bun Restaurant, Xi said "Food safety is the most important thing. People should eat at ease, which is what I am most concerned about." In 2014, during his visit in Beijing, Xi specifically asked the head of the Beijing Municipal Administration Committee about the issue of gutter oil. Likewise, during a visit to Guizhou Province, he entered a small supermarket in the parking service area and asked about the expiration date of the food. During his inspection to Inner Mongolia autonomous region, Xi exhorted the person in charge of liquid milk production base to attach great importance to food safety issues. Xi said that food safety is both a major livelihood issue and a major political issue. He stressed on many occasions that ensuring food safety is a project for people's livelihood and public support, is the responsibility of party committees and governments at all levels. "Food safety is related to the future of the Chinese nation, and the ability to give the people a satisfactory explanation on food safety is a test of our ability to govern. Whether or not the common people can eat safely and with peace of mind is already directly related to the issue of trust in the ruling party, trust in the country." In terms of food safety, General Secretary Xi Jinping has instructed that the current food safety situation in China is still serious, and the people are eager to eat with greater ease and more health. It is necessary to strengthen coordination, speed up the improvement of unified and authoritative regulatory system and system, implement the "four strictest" requirements, and ensure the food safety.

Chapter 6
Adhere to Ecological Priority and Green Development

As socialism with Chinese characteristics enters a new era, the needs to be met for the people to live a better life are increasingly broad and the demands for ecological environment are increasing. Ecological and environmental destruction has become a prominent problem in unbalanced and inadequate development. In the face of the current double challenges of easing the ecological and environmental crisis and solving the predicament of economic development, the Party Central Committee, with Comrade Xi Jinping at its core, has proposed the principle of "ecological priority" based on the objective laws of nature and realistic development needs, using forward-looking strategic thinking and vision, emphasizing the primacy and urgency of maintaining ecological benefits, making environmental resources an intrinsic element of socio-economic development, and making the "greening" of the economic activity process the main content and means of development, thus breaking through the paradox of development and protection, dialectically unifying the two, seeking a development path in ecological protection, supporting protection with green development, and pointing the way for establishing a sustainable development model with harmonious economic, environmental and social benefits.

1 Ecological Rationality is Better than Growth Rationality

A sound ecological environment is both a natural space for human survival and a source of recharge for the material resources needed for development. For a long time, China has followed the development model of "speed over quality", resulting in serious ecological deficits and becoming a "weakness" that constrains development, and it is urgent to protect the ecological environment. Against this background, General Secretary Xi Jinping proposed the principle of "ecological priority", which, as opposed to the traditional principle of economic priority, essentially insists on the fundamental status of the ecosystem and provides a criterion for judging the

J. Pan, *China's Global Vision for Ecological Civilization*,
https://doi.org/10.1007/978-981-16-4534-1_6

contradictions and conflicts between the economy, society and the environment, and has become the basic point for the construction of ecological civilization at this stage.

The report to the 18th CPC National Congress, which first mentions "prioritizing environmental conservation and protection and putting the restoration of nature first", highlights the priority status of ecological protection and shows that ecological priority mainly includes strengthening the conservation and intensive use of resources and protecting the ecological environment at source, in line with the basic state policy of "adhering to conservation of resources and protection of the environment". Since then, General Secretary Xi Jinping has repeatedly stated the concept and basic requirements of upholding ecological priority: in economic terms, priority should be given to maintaining economic efficiency, "we should consciously promote green, circular and low-carbon development, and never sacrifice the environment in exchange for momentary economic growth." In terms of institutions, it has been systematically protecting the ecological environment, and we should place greater importance on ecological environmental protection, reform the institutional mechanism for building ecological civilization on an early and pilot basis. We should improve the system of property rights and use control of natural resources capital, delineate red lines for ecological protection, and reform the management system for ecological environmental protection; in terms of ecological environmental governance, it has been coordinating resource conservation and environmental protection, "we will continue to implement sustainable development strategies, optimize the pattern of land and space development, comprehensively promote resource conservation, increase the efforts of natural ecosystems and environmental protection, focus on solving a series of problems such as smog, and strive to build a beautiful China with blue sky, green land and clean water"; in terms of social progress, maintaining the environment serves as a public good for people's well-being, we should "speed up efforts to create a good living environment in the process of pollution control and ecological restoration" and "unswervingly follow the path of ecological priority and green development", etc. The report to the 19th CPC National Congress further makes "ensuring harmony between man and nature" as the basic strategy for society with Chinese characteristics in the new era, calls for "develop spatial layouts, industrial structures, and ways of work and life that help conserve resources and protect the environment so as to restore the serenity, harmony, and beauty of nature" and points out prioritizing resource conservation and environmental protection is the fundamental premise for optimizing the spatial layout, adjusting the industrial structure and transforming the mode of development. The implementation of the principle of ecological priority cannot be achieved without institutional norms and guarantees. The 5th plenary session of the 18th CPC Central Committee has made "vigorously promote ecological progress" as the Five-year plan, which point out "building a solid ecological safety barrier, insisting on protecting the environment and promoting its natural restoration, implementing ecological protection and restoration projects for mountains, rivers, forests, fields and lakes, carrying out large-scale afforestation of the national territory, improving the natural forest protection system, and carrying out Blue Bay Initiative". We should make ecological protection to be given priority from the theoretical level to the practical level, and clarify the main contents of ecological restoration and remediation work. Opinions of the CPC Central Committee on Accelerating Ecological Progress set forth five basic principles, four key tasks and four guarantee mechanisms for the construction of ecological civilization, making the principle of ecological priority more operational and systematic, in which the content of ecological priority is subdivided into three levels. We should adhere to the basic principle of prioritizing resource conservation and environmental protection and letting nature restore itself. In terms of resource development and conservation, we should give priority to conservation and support sustainable economic and social development with minimal resource consumption; in terms of environmental protection and development, we should give high priority to environmental protection with coordinated protection and development; in terms of ecological construction and restoration, we should give priority to nature restoration and combine it with artificial restoration." Therefore, the core of adherence to the principle of ecological priority lies in coordinating the relationship between "resource development and conservation" and "ecological protection and economic development"; the key lies in meeting the requirements of "resource conservation as a priority" and "environmental protection as a priority".

From the perspective of ecological economics, the principle of ecological priority is opposite to the principle of growth priority, that is, "ecological rationality" takes precedence over growth rationality. The principle of growth priority is a one-sided pursuit of rapid economic growth and unlimited expansion of scale, taking into account only short-term economic benefits and ignoring the coordination of economic development with the carrying capacity of the ecological environment. However, ecosystems are fundamental to sustainable development systems and play a decisive role in economic and social development, as reflected on the one hand by the fact that the material resources and energy contained in ecosystems form the basic elements of economic systems and that ecological self-cycling systems provide space for production and life; and on the other hand, by the irreversible and persistent nature of ecosystem damage and the loss of ecological functions such as climate regulation, resource supply and energy cycling, which means the loss of momentum for development and space for survival.

The principle of ecological priority is a breakthrough from the traditional approach to development, in which a single economic benefit is at the core, and focuses on the multifaceted goals of coordinated economic, environmental and social development, including the triple connotation of priority of ecological law, ecological capital and ecological benefit: it gives priority to respect the balance of the ecosystem and the law of natural resource recycling, and guide social and economic activities; it give the priority to restore the ecological environment, maintain ecological functions and ensure the preservation and enhancement of resource and environmental capital; it gives priority to safeguard long-term ecological benefits and offset the loss of economic and social benefits with ecological premiums.

From the perspective of philosophy, ecological priority embodies the objective laws of nature.

Firstly, man and nature are a community of life. We must give priority to respecting the law of ecology, respecting, conforming to, and protecting nature. When Marx talks about the dual nature of man, he says that man has natural attributes. "We … all belong to and exist in nature … and are capable of recognizing and correctly applying the laws of nature"; at the same time, human beings have social attributes in that "society is the completed essence of the unity of man and nature". Human beings are part of the natural world, the development of human society depends on the natural world, and the ecological law is the objective existence before human beings and human consciousness, therefore the ecological law has the priority over the economic and social law as a fundamental and pre-requisite status, and any human activity must follow the balance of the ecosystem and the recycling law of natural resources.

Secondly, the ecological conservation paradigm requires that priority be given to promoting the preservation and enhancement of ecological capital. The traditional paradigm of industrial civilization is oriented towards maximizing economic benefits, and man and nature are in an antagonistic relationship. The development of inefficient, high consumption and high pollution solves the problem of hunger and poverty in the short term, but in the long term leads to the ecological environment carrying capacity approaching the upper limit. At present, the conditions and environment of China's economic development have undergone significant changes, and the blind pursuit of economic growth not only violates economic laws, but also exacerbates existing contradictions and brings many risks. The ecological conservation paradigm attaches importance to the value of ecosystems, maintains ecosystem productivity through ecological restoration and environmental governance, ensures the preservation and enhancement of ecological capital, takes into account the comprehensive development of economic, social and environmental benefits, and replaces the industrial civilization paradigm with the ecological civilization paradigm, which is an objective requirement for the evolution of socio-economic development from low to high levels.

Thirdly, resolving the main contradictions in development requires that priority be given to preserving ecological benefits. The materialistic dialectic holds that efforts should be focused on resolving the main contradictions that govern the development process. The limited availability of resources and the irreversible nature of environmental destruction have determined that ecological benefits are a major contradiction in development, and that ecological protection and resource conservation should be regarded as the primary task of development, with a long-term ecological premium to offset the loss of short-term economic and social benefits.

From the perspective of the practical difficulties of development, it is an urgent requirement to put ecological protection in a prominent position to break the constraints on resources and the environment, improve the quality and efficiency of economic development and improve people's livelihood in the process of development.

Firstly, upholding the principle of ecological priority is an inevitable choice to deal with the serious ecological situation. The past development model, which relied on the quantitative inputs of factors, has led to a decline in the rate of economic growth as a result of declining marginal return, with a concentration of problems such as increased energy constraints and difficult ecological and environmental protection tasks. In recent years, the frequent occurrence of pollution incidents such as smog and other pollution incidents and the decline in environmental quality have compressed the space for economic development and seriously damaged the quality people's lives, so the key to solving the economic and social crisis lies in solving the ecological crisis, and protecting the ecological advantage is protecting the development advantage.

Secondly, giving top priority to the environment is also a key factor in securing a decisive victory in building a moderately prosperous society in all respects. The report to the 19th CPC National Congress made a clear statement that "the modernization that we pursue is one characterized by harmonious coexistence between man and nature". We are in the midst of a decisive period for building a well-off society, and a high-quality ecological environment has become an important criterion for measuring the full extent of a well-off society. Only by grasping the "green" connotation of modernization, transforming the mode of economic development and increasing the efforts of ecological environmental protection can we provide more eco-products of high quality to meet the growing needs of the people for a better life.

2 Green Development is Vital to Sustainable Development

Green development advocates the integration of ecological environmental protection into all aspects of socio-economic activities, which is both a development philosophy and a development approach, providing comprehensive support for ecological protection through the promotion of "greening" reforms in ways of work and life, cultural concepts and social governance. Therefore, practicing the concept of green development has a bearing on both domestic and foreign sustainable development, which is the need for long-term transformation and development in China and the inevitable trend for future global development.

After the global economic crisis in 2008, in order to promote global economic recovery and at the same time to meet the challenges of climate change and energy crisis, under the advocacy of the United Nations and other international organizations, policy concepts such as "Green New Deal", "Green economy", "Green growth" have emerged one after another, with interconnectivity and emphasis, promoting the gradual formation of green development concepts from different perspectives.

The United Nations Environment Programme (UNEP) was the first to launch the Global Green New Deal and the Green Economy Initiative, aiming at relying on government leadership to respond to the crisis and advocating the implementation of the Green New Deal by Governments and the establishment of a low-energy, environmentally friendly and sustainable green economy growth model. UNEP has published a series of research reports on the meaning and relationship of the Green New Deal and the green economy. For example, in the *Global Green New Deal: Policy Brief*, addressing the financial, climate and food crises, it proposed that risks be addressed through two-way fiscal stimulus packages and governance policies at the international and domestic levels; in the *Towards a Green Economy*, it proposed that the green economy is a core component of the Green New Deal and defined as "an economy that enhances human's well-being and social equity while significantly reducing environmental risks and ecological scarcity"; in *The Future we Want*, it emphasized that countries should implement green economy policies according to different historical backgrounds and stages of development. Subsequently, the United Nations Economic and Social Commission for Asia and the Pacific (UNESCAP) introduced the concept of "green growth" as a prerequisite for a green economy, advocating the reduction of carbon emissions, increased resource utilization and the promotion of investment in natural capital. The Organization for Economic Cooperation and Development (OECD) issued *Towards a Green Economy*, which provides further insight into what green growth means in terms of "promoting economic growth and development while ensuring that natural assets continue to provide resources and environmental services essential for human well-being" and that economic, environmental, social, scientific and technological development should be integrated into an integrated development framework.

It can be seen that, from an international perspective, "green development" is a positive, proactive and progressive way of development, as well as a new international discourse system embodying "respecting nature, conforming to nature and protecting nature" and "social equality and justice". China's economy has now become the second largest in the world, with the gradual increase in the comprehensive strength of the country, accompanied by the sharing of international affairs and the fulfilment of the responsibilities of a major power, China should gradually change its role from following and participating in the past to leading role. The practice and dissemination of the green development concept will make "a sound ecological environment will become a growth point for people's quality of life and a launching point for presenting China's positive image", which will play an important role in building a new international dialogue mechanism.

China's "green development" is proposed in the light of the international "green wave" and in the light of the fact that China is in the decisive stage of building a well-off society and the stage characteristics of China's economy entering the new normal. It breaks through the previous single economic development goal and makes the sustainable development strategy more comprehensive and realistic.

The vision for innovative, coordinated, green, open and shared development is creatively put forward by the 5th plenary session of the 18th CPC Central Committee, which elevates green development to the level of national strategy, advocates "adhering to the basic state policy of conserving resources and protecting the environment, pushing the whole society onto a path to civilized development featuring the growth of production, an affluent life and a sound ecosystem, accelerating the development of resource-conserving and environment-friendly society, and forming a new pattern of harmonious development and modernization of man and nature", stating that the path of green development covering production, life and ecology should be followed, so as to achieve the coordination of economic, social and ecological benefits. *Opinions of the CPC Central Committee on Accelerating*

Ecological Progress put forward the "synchronization of new industrialization, informationization, urbanization, agricultural modernization and greenization". Among which, "greenization" can be described as the sublimation of green development, reflecting the dynamic process of green development to promote production, life, ecology, culture and political reform. *The CPC Central Committee Recommendations for the 13th Five-Year Plan for Economic and Social Development* extends green development to six areas: promoting harmonious coexistence between man and nature, accelerating the construction of main functional areas, promoting low-carbon recycling, conserving and utilizing resources in comprehensive and efficient manner, strengthening environmental governance and building a strong ecological security barrier. The report to 19th CPC National Congress has made "promoting green development" a key link in building a beautiful China, forming a green development mode and lifestyle and clarifying the main contents of green development in terms of green economic system, green technology innovation system, clean energy system, resource conservation and recycling, green lifestyle, etc.

In short, China's green development has experienced the sublimation from economic initiatives to political tasks, and from economic paradigm changes to deepening reforms in all areas. It has become the engine for transforming the growth model and promoting modernization, with a profound theoretical cornerstone.

First, the comprehensiveness of the sustainable development system determines that green development is a multi-dimensional comprehensive reform. The sustainable development system consists of three subsystems: economic system, social system and ecosystem. Only when the subsystems are coordinated and elements are optimally combined can economic efficiency, social equity and environmental sustainability be realized. Green development is essentially a sustainable development strategy, which means that the reform promoted by green development should cover comprehensive fields such as ways of work and life, spatial layout, industrial structure, value and political governance. This strategy takes ecological priority as the criterion of value judgment and aims at realizing sustainable economic, social and ecological development.

Second, the balance of internal circulation of the ecosystem requires that green development should adopt green, circular and low-carbon economic model. Ecosystem is a system that depends on the circular flow of material and energy, while the traditional industrial economic system is a linear economic system with one-way flow of "resources-products-wastes". This economic system breaks the balance of internal circulation of the ecosystem. Limited resources require that the economic development has to transform to the circular mode of "resources-products-renewable resources". Low-carbon economy means improving energy structure, increasing energy efficiency and reducing carbon emissions. Circular economy is to establish a sound resource and energy recycling system. Green economy is to reduce ecological and environmental damage through green and clean production and consumption. The three constitute the basic means of green development. Through ecology-based economy development, waste emissions are reduced and a virtuous circle of input and output is realized. This effectively alleviates the contradiction between unlimited demands for economic development and limited capacity of resources and environment.

Third, the valuable nature of resources and environment requires that environmental factors should be taken as an important factor of productivity in green development. Resources and environment are the main components of the ecosystem, and play a unique ecological service function, so they have ecological value. The exploitation and utilization of resources condenses the input of human labor, and the restoration and regeneration of resources and environment condenses the compensation of human labor, so they have economic value. Green development is based on the valuable nature of resources and environment, and regards environment as an important factor of productivity. Therefore, it fundamentally denies the indiscriminate exploitation of resources and the disorderly development of the environment. To establish the scientific outlook that "protecting the environment means protecting productivity", it is asked to explore new development models and driving forces with the goal of win–win development and protection.

Finally, in a practical sense, the implementation of green development is an inevitable choice to meet the internal transformation needs, ease the constraints of resources and energy, and respond to the external pressure of emission reduction. This is because:

First, from the perspective of domestic economic development stage, the implementation of green development is conducive to the reform of economic development quality, efficiency and impetus. China's economy has shifted from a stage of rapid growth to a stage of high-quality development. We are now in a critical period of transforming the growth model, improving the economic structure, and changing the drivers of growth. Taking the historical moment to push economic development to a new level, it is necessary to change the economic growth model to an intensive one. Green development nurtures new business models such as the green industry, green energy, green transportation, green architecture and green finance, which provides a feasible path for transformation and upgrading.

Second, in terms of resource and energy constraints, implementing green development is conducive to alleviating resource and energy pressure and solving the contradiction between economic development and environmental protection. China is the world's largest energy consumer and producer, and its energy consumption focuses on fossil fuels, which has caused tremendous environmental pressure. Oil and natural gas reserve-production ratios are both lower than the world average for the same period, and energy supply constraints are tightening. Water pollution and air pollution still exist. All of these greatly limit the space for development. Economic development is highly dependent on resources and energy, but the supply of resources and energy is relatively insufficient. In this case, it is necessary to use green development to establish a modern energy system, develop energy conservation and environmental protection industries, reduce the consumption of resources and energy, and alleviate environmental pollution and ecological damage. Concerning green development, it is encouraged to use technological methods to improve the utilization efficiency of resources and energy. We will establish an industrial system geared toward green, low-carbon, and circular development, and make clear the initial allocation system for energy, water, pollution and carbon permits so as to promote the economical and intensive use of resources and energy.

Third, from the view of global development, green development is conducive to implementing emission-reduction tasks and coping with the crisis of climate change. Climate change is the prominent environmental problem in the world, the biggest challenge to achieve sustainable development, and an important issue in the contemporary international community. In *The Paris Agreement*, arrangements are made for global actions to deal with climate change after 2020, and targets are set for reducing greenhouse gas emissions for all parties. The Chinese government promises to achieve the peaking of carbon dioxide emissions around 2030. These actions form external institutional constraints on China's economic development, so it is urgent to pursue a low-carbon, circular, efficient and safe path of green development characterized by "less pollution, low emission and high efficiency".

Green development integrates putting ecology first into production, life, ecology, culture and political governance. Specifically, in practice, greening ways of work should lead the formation of greening ways of life. The ways of life with green consumption as the core should force the cultivation of greening ways of work. Greening political governance and cultural values should constrain the formation of green work and life. All of these provide technological, capital, cultural and institutional support for ecological conservation.

Encouraging greening ways of production. We will encourage emerging industries to grow. These industries include industries on energy-saving and environmental protection, biotechnology, communications technology, intelligent manufacturing, advanced equipment and new energy. We will develop green finance and set up green development fund to support green and clean production. Also, we will strengthen energy consumption control in industries with high energy consumption, and implement negative lists for industries in important ecological areas. Besides, we will cultivate industrial structures and ways of production that are high in science and technology, low in resource consumption and low in environmental pollution, raise total factor production efficiency and create new areas of economic growth.

Cultivating green ways of life. We will cultivate residents' concept of environmental protection purchase and resource recycling to realize green consumption. We will accelerate energy technology innovation and develop green energy. Also, we will prioritize public transportation, promote new energy vehicles, enhance low-carbon development of transportation, and develop green transportation. We will use green building materials to improve energy-saving architecture standards and promote green architecture. We will create a good living environment during pollution control and ecological restoration.

Establishing a greening ecological system. On the premise of respecting nature, we will combine scientific planning with governance practice, draw and enforce red lines for ecological conservation, improve the layout of each region's core functions, promote agricultural development, ecological security and natural shoreline, and build ecological security shields. At the same time, we will broaden the financing channels for ecological conservation and implement major ecological projects.

Establishing greening values. We will encourage the green cultural values of being thrifty and protecting our environment in the whole society, gradually bring the values into core socialist values, and form a new social trend of advocating ecological civilization, so that the concept of "harmonious coexistence between man and nature" becomes the value standard of regulating people's behavior, and helps to implement the value of being thrifty in all aspects of production, circulation, storage and consumption.

Implementing greening social governance. We will correct the bias of evaluating political achievements only by the speed of economic growth, establish a green political performance appraisal system, strengthen the assessment of ecological environment protection from the aspects of index system, index weight and evaluation criterion, and incorporate the indicators reflecting ecological progress, such as resource consumption, environmental damage and ecological benefits, into the economic and social development evaluation system, so as to promote the reform of the system and mechanism of building an ecological civilization and protect the ecological environment with institutions.

3 Prioritizing Ecological Conservation and Boosting Green Development in Key Areas

The economy, society, and ecological environment are an organic unity that depends on each other. Once it is broken, the supply of resources for economic development will not be met; working and living spaces will be destroyed, and the people's sense of well-being will be greatly reduced. Therefore, the key to solving economic and social crises lies in solving the ecological crisis, and protecting ecological advantages means protecting development advantages. The CPC Central Committee with Comrade Xi Jinping at its core focuses on solving problems, gives priority to ecological benefits, and implements the strategy of comprehensive protection of important ecological areas such as the Yangtze River Economic Belt and Sanjiangyuan (Sanjiangyuan is an area where the Yangtze, Yellow and Mekong rivers originate). Besides, the committee advocates using ecological conservation to push the transformation of growth model, which is a respect for laws of nature and a respect for laws of the economy and society. This advocation plays a leading role in implementing putting ecology first, and has a far-reaching impact on safeguarding national environmental safety and maintaining sustainable economic development.

The Yangtze River is the golden waterway with the largest inland freight volume in the world. The Yangtze River Economic Belt relying on the golden waterway spans central, eastern and western regions in China. The economic belt covers 11 provinces and cities including Shanghai, Jiangsu, Zhejiang, Anhui, Jiangxi, Hubei, Hunan, Chongqing, Sichuan, Yunnan and Guizhou. It has an area of about 2.05 million square kilometers, with both the population and GDP exceeding 40% of the country. The economic belt is an important ecological "treasure house" in China, a vital east–west axis of territorial space development, and also a significant link between the Silk Road Economic Belt and the 21st Century Maritime Silk Road. It is the strategic underpinning for regional development. At the symposium on promoting the development of the Yangtze River Economic Belt, General Secretary Xi Jinping stressed that we would facilitate the development of the Yangtze River Economic Belt by promoting well-coordinated environmental conservation and avoiding excessive development. This embodies the essence of "prioritizing ecological conservation", and helps to drive greening reform in environment, resources, industries, transportation, and urban and rural development along the belt.

The strategic orientation of adhering to prioritizing ecological conservation in the development of the Yangtze River Economic Belt is embodied in the following five aspects. First, in terms of ecological protection, it is necessary to take the implementation of major ecological restoration projects as the priority option to promote the development of the Yangtze River Economic Belt. We will rely on reform and innovation to protect the ecology of the Yangtze River. On the premise that the eco-environmental capacity is too tight, we will make overall plans on shore and water, correctly handle the contradiction among flood prevention, navigation and power generation, and consciously promote the green, circular and low-carbon development. Second, in ensuring the economic development of the river basin, it is necessary to make overall plans for water, roads, ports, shores, production, cities, biology, wetlands and environment. We will coordinate regions' reform and development, inter-regional policies, construction in various fields and multiple resources. In this case, we can promote the coordinated development of the upper, middle and lower reaches of the Yangtze River Economic Belt, and the interaction and cooperation between the eastern, central and western regions. We can also build the Yangtze River Economic Belt into a pioneering demonstration belt, an innovation-driven belt and a coordinated development belt for China's ecological progress. Third, from the perspective of optimizing the layout of city clusters in the Yangtze River Economic Belt, we will adhere to the cooperation of large, medium and small cities, and the linkage between the eastern, central and western regions. Also, we will rely on city clusters in the Yangtze River Delta, along the middle reaches of the Yangtze River, and around Chengdu and Chongqing to facilitate the development of the Yangtze River Economic Belt. Fourth, in the selection of projects, development plans should focus on the overall strategic situation and be practical, and play a leading and constraining role. For the key directions such as protecting the ecological environment, establishing a unified market, and accelerating shifting the growth model and improving the economic structure, we will use "quick thinking" (making decisions and taking actions quickly) and do addition. For scientific utilization of water resources, optimization of industrial layout, overall planning of port and shoreline resources, and arrangement of some major investment projects, we will use "slow thinking" (making decisions and taking actions slowly) and do subtraction, and rely on scientific argumentation, comparison and selection. Fifth, in terms of the governance system, we will establish a leadership system and working mechanism of coordination and cooperation, planning guidance and market operation. We will take market and opening up as important driving forces to promote the development of the Yangtze River

Economic Belt, and see that the market plays the decisive role in resource allocation. The leading group of the Yangtze River economic belt development should give full play to their leading roles. Provinces and cities along the Yangtze River should speed up the transformation of functions of the government, improve public services, create a sound market environment, and stay ideologically and practically coordinated.

"Avoiding excessive development" doesn't mean avoiding economic development. Instead, we will rely on reform and innovation to protect the ecology of the Yangtze River. We will maintain ecological functions of the Yangtze River, and develop a green, circular and low-carbon economy by taking advantage of the superior natural conditions of the Yangtze River, which will truly bring golden benefits to the golden waterway. The carrying capacity of resources and environment is the basis for creating the strategic space of the Yangtze River Economic Belt, and the key to maintaining the balance between economic development and ecological protection. Based on this understanding, "promoting well-coordinated environmental conservation" means taking environmental protection as the basis for promoting sustainable development, forcing economic transformation and upgrading with strict constraints of the ecological environment, improving quality and efficiency, making overall plans, and enhancing systematic thinking. We must ensure that the coastal areas along the Yangtze River are linked and cooperated; government guidance and market decision are taken simultaneously, and ecological functions are coordinated with economic and social functions. We will gradually develop green, circular and low-carbon spatial configurations, industrial structures, and ways of work and life to ensure the safety of the living environment during industrialization and urbanization, which will truly bring golden benefits to the golden waterway.

We will adhere to the principle of giving priority to ecological protection and abandon the model of inefficient and blind development. Only in this way, can we gradually get rid of dependence on energy and resource consumption, explore new ways with increased production and further improvement in the environment, and build the Yangtze River Economic Belt into a golden economic belt featuring more beautiful ecology, more smooth transport, more coordinated economy, more consistent market management and more scientific mechanisms.

When inspecting the Qinghai Province, General Secretary Xi Jinping emphasized that Qinghai holds an important and special ecological position, and must shoulder the great responsibility of protecting Sanjiangyuan, known as China's "Water Tower". It is necessary to prioritize environmental protection, combine natural recovery of the environment with artificial restoration, implement the planning requirements of main functional areas according to the actual circumstances, and comprehensively strengthen the main functions of safeguarding national ecological security. It is necessary to promote ecological projects, energy conservation and emission reduction, environmental improvement, and building of beautiful cities and countryside. We will strengthen construction of nature reserves, carry out the pilot project of Sanjiangyuan national park system, enhance ecological protection around the Qinghai Lake, and better prevent and control desertification. Also, we will strengthen the construction of alpine grassland, turn more grazing land and marginal farmland into forests and grasslands, and implement the three-north shelter forestation project (a large-scale artificial forestry ecological project built in the three north regions—northwest, north and northeast of China). Besides, we will enhance energy conservation and emission reduction, and comprehensive environmental management so as to guarantee that "rivers of clean water run eastward".

The Sanjiangyuan area, where the Yangtze, Yellow and Mekong rivers originate, is located in the hinterland of the Qinghai-Tibet Plateau and in the south of Qinghai Province. It is known as China's "Water Tower", and is the lifeblood of fresh water supply in China. Since the Sanjiangyuan area is of concentrated biodiversity and sensitive ecological environments, to protect the local ecological environment is related to the ecological, economic and social benefits of the western region and even the whole country. The ecological benefits are embodied in the ecological functions of climate regulation, water conservation, conservation of water and soil, and maintenance of biodiversity. The economic benefits are reflected in providing biological resources and bearing space necessary for work and life. The social benefits are shown in the Sanjiangyuan area being located at the junction of Qinghai, Sichuan, Gansu and Tibet, and taking on the responsibility of protecting nomadic culture and religious culture and maintaining ethnic unity and social stability. Therefore, restricting or prohibiting development in the Sanjiangyuan area is not hindering development, but exploring new development opportunities and obtaining triple benefits through ecological protection.

A series of institutional innovations, such as "an integrated pilot zone for the country's ecological protection", "project planning for ecological conservation and construction of Sanjiangyuan" and "a national park system", have provided policy support for ecological protection in the Sanjiangyuan area. The approval of the master plan on an integrated pilot zone for the country's ecological protection indicates that ecological protection of Sanjiangyuan has risen to the status of national strategy. In the second-stage project planning for ecological conservation and construction of Sanjiangyuan in Qinghai Province, it has expanded the scope of governance of Sanjiangyuan from 152,300 to 395,000 km^2. Also, a number of major ecological projects are deployed, such as promoting ecological protection of Sanjiangyuan in Qinghai, building an integrated pilot zone for national ecological security shields in Gansu Province, controlling the factors causing sandstorms in Beijing and Tianjin, and carrying out environmental governance of lake water in the Great Lakes of China. The pilot project of the Sanjiangyuan national park system covers three parks of headwaters of Yangtze, Yellow and Mekong rivers, with a total area of 123,100 km^2, accounting for 31.16% of the area of Sanjiangyuan. The purpose is to build an ecological protection management system with confirmed ownership, clarified rights and responsibility and effective supervision, and realize the systematic protection of natural ecological spaces such as mountains, water, forests, grass and lakes in the typical and representative areas of Sanjiangyuan. The purpose is also to build the Sanjiangyuan national park into a demonstration area for ecological protection and restoration on the Qinghai-Tibet Plateau, a pioneering area for joint contribution, shared benefits and harmonious coexistence between man and nature in Sanjiangyuan, and an area of natural protection exhibition and ecological culture inheritance on the Qinghai-Tibet Plateau.

Under the national top-level design, ecological protection in the Sanjiangyuan area has been carried out in an orderly manner, and the expected triple benefits will gradually appear. In terms of ecological benefits, after ten years of the first-stage ecological protection projects, the macro-structure of the ecosystem in the Sanjiangyuan area has been partially improved. Grassland degradation has been initially curbed. The contradiction between grass and livestock has been alleviated. The ecological function of wetlands has been gradually improved. The lake water area has been significantly expanded. The water conservation and water supply capacity of the basin have been greatly enhanced. The vegetation coverage rate in severely degraded areas has been significantly improved. The ecology in key governance areas has been improved, which is mainly manifested in increasing vegetation coverage, water resources and biodiversity. General Secretary Xi Jinping pointed out: "To protect the ecological environment, we must first get specific and real-time information about its situation, and well build and use the ecological environment monitoring network". Departments on environmental protection, agriculture and animal husbandry, forestry, science and technology, water conservancy, and meteorology have joined forces to build an ecological monitoring, assessment and early warning system of Sanjiangyuan, ensuring the comprehensive monitoring of forests, grass and water, enhancing the ecological governance capacity, and guaranteeing the ecological security of our country. As far as economic benefits are concerned, in the process of ecological and environmental protection, we will rely on scientific and technological innovation to continuously enhance regional management, and improve production environment and productivity. Measures are taken to improve the regional economic development and broaden income sources of farmers and herdsmen. These measures include rural energy development, ecological animal husbandry infrastructure, forest seedling base construction and systematic protection of aweto resources. With regard to social benefits, along with the promotion of ecological protection in the Sanjiangyuan area, there is also a political view of ecological people's livelihood for poverty reduction and alleviation. General Secretary Xi Jinping pointed out: "Protection of Sanjiangyuan is a major policy made by the Party Central Committee, and ecological migration is essential for this policy, so it should be properly organized and implemented". We must implement poverty reduction and alleviation policies, protect the scarce Tibetan and pastoral culture, continuously improve people's living standards, and promote social harmony.

4 Innovating the Realization Mechanism of Prioritizing Ecological Conservation and Boosting Green Development

Prioritizing ecological conservation and boosting green development complement each other. Prioritizing ecological conservation creates conditions for boosting green development, and boosting green development provides support for prioritizing ecological conservation. The relationship between prioritizing ecological conservation and boosting green development is essentially the coordination between protection and development, which requires a benefit coordination mechanism to realize the unity of ecological benefits, economic benefits and social benefits. When the economic benefits and social benefits lag behind the ecological benefits, we can rely on advance compensation realized by means of transfer payment and preferential policies, use ecological conservation to push the transformation of growth model, and utilize ecological advantages to create development opportunities. When the economic benefits and ecological benefits are in conflict and cannot be coordinated, they should be chosen according to the principle of prioritizing ecological conservation. The economic benefits and social benefits should be abandoned when necessary, so as to obtain long-term green incentives and make up for the loss of short-term benefits.

Specifically, prioritizing ecological conservation takes the protection of ecological environment as the primary task, which is the premise and criterion of development. It advocates respecting nature, following its ways, and protecting it, and holds that protecting ecological advantages means protecting development advantages. In essence, it aims to pursue the harmonious development between man and nature, adapt work and life to the carrying capacity of resources and environment, and maintain ecological balance and stability of ecosystem functions, thus creating conditions for boosting green development. Green development integrates the criterion of prioritizing ecological conservation into production, life, culture and political governance in an all-round way, which helps to realize ecological and environmental protection. The foothold of green development lies in development. "Green" represents the direction of development, that is, following a low-carbon, clean and efficient growth model. "Development" is a comprehensive reform covering production, life, culture and ecology. It tries to adapt social and economic activities to the carrying capacity of resources and environment through the transformation of growth model, and attaches importance to the quality and effect of development, and well-rounded human development.

Prioritizing ecological conservation and boosting green development creatively realize the dialectical unity of protection and development. This is a major theoretical breakthrough in the ecological progress in China, and its theoretical value is reflected in three aspects.

First, prioritizing ecological conservation and boosting green development are the latest achievements in adapting Marxist ecological view and historical materialism to the Chinese context, which promote the continuous improvement of the system of building an ecological civilization. The emphasis of prioritizing ecological conservation and boosting green development on ecological protection not only inherits the concept of "giving priority to the external nature" advocated by Marxist ecological view, but also realizes adapting Marxist historical materialism to the Chinese context. Marxist historical materialism holds that "the means of production in material life restrict the process of overall social life, political life and spiritual life", and the change of means of production plays the decisive role in promoting social and historical progress. Prioritizing ecological conservation and boosting green development will bring unprecedented green changes in means of production and life by injecting the principle of putting ecological protection first into development, which helps to better meet the intensive and efficient requirements of sustainable development and contribute to the overall progress of society. This concept effectively links Marxist ecological view with historical materialism, and is innovated in combination with China's development needs, indicating that the focus of building an ecological civilization system lies in putting ecological protection in a prominent position. The key areas are not only the governance of ecological environment, but also the comprehensive reform of economy, politics and culture.

Second, prioritizing ecological conservation and boosting green development are the achievements of Chinese traditional ecological thoughts, which shape the ecological and cultural values of harmonious coexistence between man and nature. General Secretary Xi Jinping pointed out: "To resolve various contradictions between man and nature, between man and man, and between man and society, we must rely on the influence, education and encouragement of culture." There are sufficient ecological thoughts of harmony and unity between man and nature in Chinese traditional culture. In the *Book of Changes*, it is first proposed "the three powers of heaven, earth, and man", and the concept that "the laws governing the ways of heaven are yin and yang". It also holds that the way of humans should learn from the way of heaven and the way of earth. Taoism advocates that "Dao operates naturally". Confucianism holds that "heaven and man are united as one". All of these indicate the philosophy of harmony and unity between man and nature, and respecting nature. Prioritizing ecological conservation and boosting green development emphasize that protection and development are inseparable, which is the contemporary achievement of ancient naive materialism of "respecting nature and following its ways". It will form an ecological culture of harmonious coexistence between man and nature in the whole society, and guide people to consciously cultivate green, frugal, low-carbon and healthy awareness.

Third, prioritizing ecological conservation and boosting green development provide guidelines for balancing ecological and economic benefits. In the past, ecological civilization emphasized "don't pursue economic growth at expense of environment", but the emphasis on ecological protection was insufficient. When there is a contradiction between economic benefits and ecological benefits, the selection criteria are vague. The principle of prioritizing ecological conservation has unprecedentedly promoted the national strategic position of building an ecological civilization. It clearly defines the criteria that laws of ecology take precedence over laws of economy, and ecological benefits take precedence over economic benefits. It also promotes the development model transformation from putting economy first to putting ecology first through green development. Adhering to prioritizing ecological conservation does not mean that we don't pursue economic growth. Instead, we will abandon the traditional model of "high input, high pollution, high consumption and low efficiency", and pursue a green, circular and low-carbon model. We will pay attention to economic growth, environmental protection and well-rounded human development. Adhering to prioritizing ecological conservation does not mean sacrificing economic benefits. Instead, it advocates adopting a green economy model that fully takes environmental factors into consideration, and realizing compensation for economic benefits by bringing dividends such as "optimization of the economic structure, improvement of the ecological environment and improvement of the people's wellbeing".

Based on the correlation between prioritizing ecological conservation and boosting green development, to implement prioritizing ecological conservation and form a green development model, we need to rely on mechanism innovation in science and technology, concepts, culture and systems to promote the transformation of ecological advantages to development advantages.

We will rely on scientific and technological innovation to foster new business models and provide new means of governance. We will create a market-oriented system for green technological innovation, and develop a market-oriented system for technological innovation in which enterprises are the main players and synergy is created through the joint efforts of enterprises, universities, and research institutes. We will support innovation by small and medium-sized enterprises and encourage the application of advances in science and technology. With the help of green finance, we will develop energy-saving and environmental protection industries, clean production industries and clean energy industries, and promote cleanness of production and diversification of industrial structures. Meanwhile, we will improve energy-saving technology, resource recycling technology, and development and utilization technology of new energy and materials. Also, we will explore new measures of emission reduction and pollution control, and enhance the capacity of ecological and environmental governance.

We will rely on development paradigm innovation to promote new industrialization and urbanization. In the past, European and American countries follow the industrialized path of developing the economy at the cost of the environment, and the urbanized path of destruction before renovation. This is essentially a utilitarian-oriented development paradigm of industrial civilization, leading to serious problems such as resources waste, environmental pollution, ecological damage and social problems that arose with urban development. China has to avoid repeating their mistakes. China's industrialization and urbanization should be transformed into the ecological-oriented development paradigm of ecological civilization, so as to avoid the restoration of ecological damage afterwards and reduce the development cost as much as possible. We will keep a balance between projects on social and economic development and those on ecological protection. Efforts should be redoubled to enforce ecological conservation red lines, designate permanent basic cropland, and delineate boundaries for urban development. It is necessary to pursue green industries, transportation, architecture and communities, and build ecological corridors and biodiversity protection networks. We will also improve urban green infrastructure services, and create a green, low-carbon, ecological and livable environment for development.

We will rely on cultural innovation to cultivate a new trend of ecological civilization. "The three powers of heaven, earth, and man" proposed in the *Book of Changes*, "Dao operates naturally" advocated by Taoism, and "Heaven and man are united as one" held by Confucianism form the cultural genes of ecological civilization. Prioritizing ecological conservation and boosting green development emphasize respecting nature and following its ways, which is the achievement of the traditional ecological view in the contemporary era. It is necessary to cultivate ecological culture, promote the mainstream values of ecological civilization in the whole society, and guide people to internalize the value standards of ecological protection in the mind and externalize them into practice. We will create a social trend of being thrifty, shape a good national image of "civilized China and beautiful China", and then strengthen our cultural soft power.

We will rely on institutional innovation. While implementing the strictest environmental protection system, we will improve the compensation function of resource wealth, and the transformation to social cost. The innovation of a series of systems helps to attach importance to the value of resources and environment, and clarify the standards for the protection and development of ecological environment. These systems include the property rights trading system for natural resources assets, the ecological compensation system, the management system for state-owned natural resource assets, the green national wealth accounting system, and the system for ecological conservation red lines. On the one hand, we will use resource wealth to implement ecological compensation, make up for the depletion of resource value, and improve the compensation for resource damage; on the other hand, we will guide the transformation of natural resource wealth into social capital and invest it in social development, so as to realize the sharing of resource benefits among all people and generations.

Ecological problems are essentially development problems. We will adhere to the development philosophy of "harmonious coexistence between man and nature", follow the "green, circular and low-carbon" growth model, and set up the development goal of "sustainable economy, politics, society and ecology". The path of green development, guided by the principle of prioritizing ecological conservation, is to integrate development and protection, transform ecological advantages into development advantages, and finally make lucid waters and lush mountains bring great benefits.

Chapter 7
Providing Legal Guarantee and Institutional Support for Building an Ecological Civilization

The 19th CPC National Congress focuses on answering the profound question that socialist modernization with Chinese characteristics has entered a new era. In answering this question, our Party has been guided by Marxism-Leninism, Mao Zedong Thought, Deng Xiaoping Theory, the Theory of Three Represents, and the Scientific Outlook on Development. It has continued to free our minds, seek truth from facts, move with the times, and apply a realistic and pragmatic approach. In answering this question, our Party has continued to uphold dialectical and historical materialism, has considered carefully the new conditions of the era and the new requirements of practice, and has adopted an entirely new perspective to deepen its understanding of the laws that underlie governance by a Communist party, the development of socialism, and the evolution of human society. It has worked hard to undertake theoretical explorations, and has achieved major theoretical innovations, ultimately giving shape to the Thought on Socialism with Chinese Characteristics for a New Era. The Thought on Socialism with Chinese Characteristics for a New Era makes clear that the overall goal of deepening reform in every field is to improve and develop the system of socialism with Chinese characteristics and modernize China's system and capacity for governance. It also makes clear that the overall goal of comprehensively advancing law-based governance is to establish a system of socialist rule of law with Chinese characteristics and build a country of socialist rule of law.

Based on this, in the third part of the 19th CPC National Congress report, that is, the ninth part of the "Thought on Socialism with Chinese Characteristics for a New Era and the Basic Policy", it is pointed out that we must realize that lucid waters and lush mountains are invaluable assets and act on this understanding, implement our fundamental national policy of conserving resources and protecting the environment, and cherish the environment as we cherish our own lives. We will adopt a holistic approach to conserving our mountains, rivers, forests, farmlands, lakes, and grasslands, and implement the strictest possible systems for environmental protection. In the ninth part—Speeding up Reform of the System for Developing an Ecological Civilization, and Building a Beautiful China—of the 19th CPC National Congress

J. Pan, *China's Global Vision for Ecological Civilization*,
https://doi.org/10.1007/978-981-16-4534-1_7

report, it is specifically pointed out that we will step up efforts to establish a legal and policy framework that promotes green production and consumption, and promote a sound economic structure that facilitates green, low-carbon, and circular development; we will improve environmental management systems, and, in a unified way, perform the duties of the owner of public-owned natural resource assets, the duties of regulating the use of all territorial space and protecting and restoring ecosystems, and the duties of monitoring the discharge of various pollutants in urban and rural areas and conducting government enforcement of laws. We will establish systems for developing and protecting territorial space, improve supporting policies on functional zones, and develop a nature reserves system composed mainly of national parks. We will take tough steps to stop and punish all activities that damage the environment.

Building an ecological civilization is a profound change involving ways of work, life and thinking, and values. To achieve such a fundamental change, we must rely on institutions and rule of law. Some outstanding problems in China's environmental protection are mostly related to imperfect systems, poor mechanisms, and incomplete rule of law. General Secretary Xi Jinping pointed out: "Only by making the best institutional arrangements and implementing the strictest rule of law can we provide reliable guarantee for building an ecological civilization." We should attach great importance to the strict binding role of institutions and rule of law in building an ecological civilization. With the spirit of reform and innovation, and greater political courage and wisdom, we should lose no time to deepen the reform of the system for developing an ecological civilization, and resolutely get rid of all the ideological concepts and institutional defects that hinder the building of an ecological civilization. It is necessary to establish a systematic and complete system to protect the ecological environment with the system. Also, we must achieve sound lawmaking, strict law enforcement, impartial administration of justice, and the observance of law by everyone, and promote the modernization of China's system and capacity for governance.

1 Implementing the Strictest System for Ecological and Environmental Protection with the Rule of Law

The rule of law is an important guarantee for a country's development and a basic way of governance. For the first time, the Fourth Plenary Session of the Eighteenth CPC Central Committee, in the form of a plenary session, discussed and deployed how to comprehensively advance law-based governance. It requires implementing Chinese socialist rule of law-based theories, and forming a complete legal norm system, an efficient implementation system of rule of law, a strict supervision system of rule of law, and a strong guarantee system of rule of law, so as to achieve sound lawmaking, strict law enforcement, impartial administration of justice, and the observance of law by everyone, and promote the modernization of China's system and capacity for governance. General Secretary Xi Jinping emphasized that we must "strive to build rule of law in China to give full play to the role of the rule of law in China's governance and social management", and "comprehensively advance sound lawmaking, strict law enforcement, impartial administration of justice, and the observance of law by everyone, pursue coordinated progress in law-based governance, law-based exercise of state power, and law-based government administration, promote the integrated development of rule of law for the country, the government, and society, and continue to break new ground in law-based governance".

Ecological civilization must also rely on the rule of law to achieve the modernization of China's system and capacity for governance. The Fourth Plenary Session of the Eighteenth CPC Central Committee put forward: "Protecting the ecological environment with a strict legal system, accelerating the establishment of a legal ecological civilization system that effectively restricts development behaviors and promotes green, circular and low-carbon development, strengthening producers' legal responsibilities for environmental protection, and substantially increasing the cost for breaking the law". It's also suggested that we must establish a legal property rights system for natural resource assets, improve the legal system for developing and protecting territorial space, and formulate and improve laws and regulations on ecological compensation, prevention and control of soil, water and air pollution, and marine ecosystem conservation, ultimately promoting the ecological progress. Based on this.

First, sound lawmaking is the premise. General Secretary Xi Jinping pointed out: "There is no end to practice and legislation. The task of improving the socialist legal system with Chinese characteristics is still tough." The same is true of legislation in ecological civilization. With the deepening of ecological civilization, China's current ecological protection laws and regulations cannot fully meet the urgent needs of our ecological and environmental protection. For example, concepts and guiding ideologies for lawmaking are outdated. There are some legislative gaps in the current environmental resources' legislation; supporting laws and regulations are not timely formulated, and other environmental management means lack legal basis. Some regulations have not met the needs of economic and social development, thereby lacking timeliness. Some laws are not well connected between the former law and the latter law, and some are not consistent with relevant laws and regulations, which brings certain difficulties to the determination of environmental responsibilities. Besides, several laws and regulations are too abstract and not easy to be implemented effectively. Therefore, the task of developing ecological legislative systems in China is still arduous amid the complex and fast-changing international economic situation, which puts forward a series of new tasks and questions for developing ecological legislative systems. Strengthening legislation has become the top priority in developing legislative systems for ecological civilization. First of all, the traditional legislation's position of taking rights as the starting point or the awareness of legal protection based on rights should be totally discarded. Fundamentally speaking, according to the ecological civilization concept of respecting nature, following its ways, and protecting it, ecological legislation must be bound by ecological laws and can be only compiled within the scope permitted by natural laws. Legislators should learn to make their will obey the laws of nature, consciously take ecological laws as the criterion for making laws, and pay attention to using natural laws to check whether norms and systems produced through legislative procedures are correct or not. If legislative activities are often accompanied by balance and coordination, then legislation under the condition of ecological civilization should first coordinate the relationship between human's usual activities of exploiting nature and ecological protection, instead of just balancing social relations such as nationalities, political parties, central and local governments, and overall and local situations. Next, under the guidance of "socialist ecological civilization" put forward by the 19th CPC National Congress, we will promote the development of environment acts to ecological acts, and gradually realize China's ecology-based environment acts. We will realize the transformation of the legislative focus from the current "economic priority" to "coordination between ecological protection and economic development". Also, we will advocate the adaptation between population and ecology, and the adaptation between economy and ecology. All special laws under the Basic Law of Environment should embody this spirit in legislative purposes, principles and content. The concept of ecological civilization should also be incorporated into the Chinese Criminal Law, Civil and Commercial Law, Administrative Law, Economic Law, Procedure Law and other relevant laws to promote making relevant laws on the basis of ecology. Finally, carrying out lawmaking in a well-conceived and democratic manner is the fundamental way to improve the quality of legislation. The core of sound lawmaking lies in respecting and embodying objective laws, and the focus of democratic lawmaking is about serving the people and relying on the people. We should improve the mechanism of sound and democratic lawmaking, innovate the way of public participation in legislation, and take into account opinions from a wide range of our society.

Second, strict law enforcement is the key. General Secretary Xi Jinping pointed out: "The very life and authority of the law lie in its implementation. 'In the governance of a country, it is the enforcement, not the enactment, of law that presents the greatest challenge.' No amount of legislation can be of any use if laws are not enforced, shelved, or poorly enforced or superficial. The key of comprehensively advancing law-based governance should be to ensure strict law enforcement, so that 'once enacted, a law must be enforced and anyone who violates it must be held to account'."

Environmental law enforcement is one of the important means to ensure the safety of ecological environment. Due to diverse historical and practical reasons, there are still multiple problems and difficulties in China's administrative law enforcement in environmental protection. Some local officials are weak on awareness of environmental protection and law, and lack urgency in protecting the environment. They even counterpose environmental protection to economic development, and emphasized "developing the economy at the cost of the environment", "getting on the bus before buying tickets", and "handling special cases with special methods". In some areas, "local policies" and "local regulations" were introduced in the name of the governments, which explicitly restrict law-based administration of environmental protection departments, blatantly protect illegal activities, and set obstacles to environmental law enforcement and supervision and management, resulting in many "special" enterprises remaining aloof from environmental supervision for a long time. The environmental pollution in the areas under their jurisdiction has not been improved, and environmental disputes have continued. Some enterprises even violently resist the law. In some regions, environmental protection supervision is weak, and even local protectionism exists. In several regions, environmental standards are not implemented, and projects that causes serious pollution to the environment are approved in violation of rules and laws. Some local officials can't make up their minds to close polluting factories, can't take effective measures, and even turn a blind eye. In other regions, environmental law enforcement is hindered, which makes the environmental supervision of some parks and enterprises out of control. If this situation is not changed, ecological legislation will be a castle in the air.

Third, impartial administration of justice is the guarantee. General Secretary Xi Jinping pointed out: "Administration of justice is the last line of defense to uphold social fairness and justice. I once quote a passage from British philosopher Bacon, who said: 'An unjust trial is worse than ten crimes. A crime is a contempt or violation of the law just as a current is polluted, but an unjust trial damages the law just as headwaters are polluted.' There is some profound truth on it. If the line of defense on administration of justice lacks credibility, social justice will be generally questioned, and social harmony and stability will be difficult to be realized."

What we know as environmental administration of justice, from a broad perspective, is a general term for judicial activities related to the environment. At present, the universal problems faced by environmental administration of justice are highlighted in four aspects: First, it is difficult to obtain evidence for cases involving environmental protection, to identify the limitation period of a lawsuit, to apply laws and to enforce rulings. Cases involving environmental protection are generally of the cross-regional and cross-sectoral characteristics, and the lag of harm results and the lack of relevant legal basis bring about the difficulties above. Second, the appraisal institutions, qualifications and procedures of cases involving environmental protection need to be standardized. Third, there is a lack of effective cooperation between the departments in charge of environmental resources and the judicial departments, and judicial means and administrative means are difficult to be connected, resulting in a large number of cases that destroy environmental resources not being settled in accordance with judicial proceedings. Fourth, the awareness of the people's courts to strengthen environmental judicial protection needs to be strengthened; the trial power of cases involving the environment is insufficient, and the filing, jurisdiction and judicial statistics of related cases need to be standardized.

Public interest lawsuits have been formed in ancient Rome. Compared with private interest lawsuits, public interest lawsuits are lawsuits to protect public interests, which can be brought by any citizen except for special provisions in law. Since the mid-twentieth century, the increasingly serious environmental problems and the rising environmental protection movements have made the environmental right, as one of personal rights, attract public attention. Therefore, the activeness of citizens' environmental litigation is also a sign of judging the implementation of environmental law. In the United States, in order to encourage citizens' environmental litigation, it is stipulated in the Clean Water Act in the U.S. that after the prosecutor wins the case, the losing party shall bear all the expenses incurred by the prosecutor, and the state shall reward the prosecutor again. According to the River and Harbor Act in the U.S., a prosecutor who brings a lawsuit against an environmental offender can get a part of the fine. Regarding this, only according to Article 108 of Civil Procedure Law of the People's Republic of China, "the plaintiff must be a citizen, legal person or any other organization that has a direct interest in the case", the conditions for determining the subject qualification of environmental lawsuits have already involved the determination of the subject qualification of public interest lawsuits. General Secretary Xi Jinping specifically pointed out: "In real life, for cases involving infringement or danger to the public interests of the state and society, such as ecological environment and resource protection, caused by the illegal exercise of power or dereliction of duty of some administrative organs, these organs are not permitted to initiate public interest lawsuits, for they have no direct interest with citizens, legal persons and other social organizations. This results in a lack of effective judicial supervision of illegal administrative actions, which is not conducive to promoting law-based government administration and strict law enforcement, and strengthening the protection of public interests. Initiating public interest lawsuits by procuratorial organs is conducive to optimizing the allocation of judicial powers, improving administrative litigation systems, and promoting the development of rule of law for the government."

Fourth, the observance of law by everyone is the foundation. General Secretary Xi Jinping pointed out: "The authority of the law comes from the people's heartfelt support for and faith in it. People's rights and interests need to be safeguarded by the law, and the authority of the law needs to be upheld by the people." We must promote the socialist rule of law, so that all our people will faithfully advocate, consciously observe, and firmly safeguard the socialist rule of law. Confucius said, "If people are guided by governmental decree and made to behave themselves through punishments, they will avoid punishment, but will have no sense of shame. If they are guided by morality and behave themselves in accordance with social norms, they will have a sense of shame and will follow rules." Ecological environment is the fairest public product, and the most inclusive wellbeing of people. Survival, development and integration into nature of everyone living on the earth are all related to the environment. In terms of Chinese culture, ecological culture has always been the core of traditional culture, which embodies the mainstream spirit of Chinese civilization. Chinese Confucianism puts forward "heaven and man are united as one", and Chinese Taoism advocates "Dao operates naturally". Throughout the dynasties, there are clear regulations and prohibitions on environmental protection. The descendants of the Chinese nation also regard ecological awareness as the mainstream awareness of guarding Chinese traditional culture for thousands of years. In this sense, the observance of law by everyone is consistent with building an ecological civilization by the whole people.

2 Strengthening Systemic and Institutional Building of an Ecological Civilization

Institutional building of an ecological civilization is the key and decisive achievement of deepening reform of the system for developing an ecological civilization. In the 19th CPC National Congress, it is pointed out that the period between now and 2020 will be decisive in finishing the building of a moderately prosperous society in all respects. We must follow the requirements set out at our 16th, 17th, and 18th National Congresses on building a moderately prosperous society in all respects, act in response to the evolution of the principal contradiction in Chinese society, and promote coordinated economic, political, cultural, social, and ecological advancement. We must constantly clarify the relationships between the Beautiful China initiative and the system for developing an ecological civilization, and between the system for developing an ecological civilization and the institution of ecological civilization from the following bases: the indispensable requirement that we must finish the building of a moderately prosperous society in all respects, then build China into a modern socialist country that is prosperous, strong, democratic, culturally advanced, harmonious, and beautiful, and realize the Chinese Dream of national rejuvenation; the principle of deepening reform in every field at a new historical starting point, and the strategic position that deepening the reform of the system for developing an ecological civilization is an important component of deepening reform in every field. The basic logic is: First, comprehensively deepening reform of the system for developing an ecological civilization is an important component and new measure for deepening reform in every field. Second, institutional building is the focus of reform of the system for developing an ecological civilization. Third, systemic and institutional building is a systematic measure for institutional building of an ecological civilization. Fourth, the formation of a more mature and stereotyped institution of an ecological civilization is a decisive achievement in reform of key links. Therefore, to build an ecological civilization, we must establish a systematic and complete system for building it, and protect the ecological environment with the system. It is necessary to improve the property rights system and use regulation system for natural resources assets, draw red lines for the protection of the ecosystems, implement the system for paying for resource consumption and compensating for ecological damage, and reform the management system of ecological and environmental protection.

First, improving the management system of national natural resource assets. There is a difference between the state exercising ownership and management of public-owned natural resource assets and the state exercising supervision over natural resources within the national territory. The former is the right in the sense of an owner, while the latter is the power in the sense of a manager. This requires the improvement of the supervision system for natural resources, and the fulfillment of our duties as the managers of our territorial space. The owners of state-owned natural resource assets and the managers of state natural resources must act independently, while cooperating with and supervising each other. General Secretary Xi Jinping pointed out: "Improving the management system of national natural resource assets is a major reform to improve the property rights system for natural resources assets, and an inherent requirement for establishing a systematic and complete system for building an ecological civilization." Some outstanding problems in China's ecological and environmental protection are related to imperfect systems to a certain extent. One of the reasons is that owners of public-owned natural resource assets are vague, and the rights and interests of the owners are not implemented. From the reality, building an ecological civilization is a systematic project, but the localized and compartmentalized management system leads to the common phenomena of conflicting policies from different departments, layers of duplication in pollution control, and too many government departments being involved in water management. In terms of central or local fiscal support or departmental interests, there are overlapping and competing powers. However, as for supervision or administrative accountability, there is often a supervision vacuum that many officials are in charge, but none of them assume responsibilities. Regarding this, General Secretary Xi Jinping stressed: "To control the exploitation of natural resources and restore the ecosystem, we must follow the laws of nature. If people only tend to their own responsibilities, for example, growing trees, regulating rivers or protecting farmlands in isolation, they are prone to gaining in one area and losing in another, which eventually leads to systemic destruction of the ecology." In the 19th CPC National Congress report, General Secretary Xi Jinping further put forward that "we will adopt a holistic approach to conserving our mountains, rivers, forests, farmlands, lakes, and grasslands". The essence of adopting a holistic approach to conserving our mountains, rivers, forests, farmlands, lakes, and grasslands requires a multi-pronged approach to promote the ecological progress with systematic mentality. We should consciously think beyond our own self-interest, further establish and improve the strict supervision system for ecological protection from the top-level design, integrate all natural ecosystems such as grasslands, forests, wetlands, oceans and rivers, as well as all protected areas such as nature reserves, forest parks and geological parks, and implement scientific, effective and integrated management to gradually recuperate the overdrawn resources and environment. We will expand green ecological space such as forests, lakes and wetlands, draw and enforce red lines for the rational resource utilization, the quality of the environment, and the protection of the ecosystems in a comprehensive and systematic manner, and gradually enhance environmental capacity through the holistic approach to conserving our mountains, rivers, forests, farmlands, lakes, and grasslands.

Second, establishing a management system of the resources, ecology and environment. General Secretary Xi Jinping pointed out: "In terms of systems, we should establish a management system of the resources, ecology and environment, accelerate the establishment of the system for territorial space development and protection, and strengthen the prevention and control system of water, air and soil pollution. Also, we should establish a system for paying for resource consumption and compensating for ecological damage—a system that responds to market supply and demand and resource scarcity, recognizes ecological values and requires compensation in the interests of later generations. We should improve the system of accountability for ecological and environmental protection and the system of compensation for environmental damage. All the systems are conducive to strengthening institutional constraints."

Accelerating the establishment of the system for territorial space development and protection. It is through land use that ecological progress can be advanced. To make overall plans for top-level design from key aspects, we must first design the framework of territorial space development. In accordance with the principle of balancing population, resources and environment, and unifying economic, social and ecological benefits, we should make overall plans for territorial space development, and coordinate population distribution, economic layout, territorial use, and ecological and environmental protection. We should also scientifically lay out production space, leave more restoration space for nature, leave more fertile land for agriculture, and leave beautiful homes with blue skies, pollution-free land and clear waters for future generations. It is necessary to adapt to each region's core functions, strictly implement environmental functional zonings, seriously follow the main functional orientation of optimized development, key development, restricted development and prohibited development, and draw and enforce red lines for the protection of the ecosystems in important ecological areas, and sensitive and fragile areas of land and marine ecological environment. Furthermore, we should build scientific and reasonable patterns of urbanization promotion, agricultural development and ecological security to ensure national and regional ecological security and enhance ecological services. The concept of ecological red lines must be firmly established. "We should study and demonstrate what should be included in the ecological red lines, how to protect the ecological red lines based on systems, as well as how to protect the good ecosystems as much as possible. After all items are included, the entire Party and all the Chinese people should never cross the red lines for the protection of the ecosystems. As for ecological conservation, we are not permitted to cross the line, otherwise we shall be punished."

Establishing a system for paying for resource consumption. We will reform the unreasonable system for setting prices for resource products, so that the resource prices are capable of reflecting supply and demand in the market, resource scarcity, and the cost of environmental damage. We will gradually extend resource taxes to cover more types of resources, increase the fees, and implement tax calculation methods that are conducive to resource conservation. Also, we will levy taxes on ecological and environmental protection in a timely manner, and reasonably raise pollutant discharge fees. We will continue to restrict the export of raw materials, rough machining and products with high energy consumption, excessive packaging and high pollution.

Establishing a system for compensating for ecological damage. According to the principle of "who develops protects, who destroys restores, and who benefits compensates", we will strengthen the policy of paying for resource consumption and pollution, and comprehensively apply economic means such as the price, taxation, finance, industry and trade so as to change the present situation of low cost of resource consumption and environment deterioration, and form a scientific and reasonable compensation mechanism, investment mechanism, and property rights and use rights trading mechanism of resources and environment. In doing so, we will fundamentally solve the contradiction between economy and environment, and development and protection.

Third, establishing a system of accountability. General Secretary Xi Jinping pointed out: "Resources and environment are public goods, and those who damage and destroy resources and environment must be held to account", and "We should establish a system of accountability. Those, who make blind decisions and lead to serious consequences regardless of the ecological environment, must be held to account and should be held accountable for life". At present, the overall deterioration of China's environment has not been fundamentally contained, and major environmental incidents occur frequently, with increasing environmental risks, which seriously threatens the safety of people's lives and property. However, after a major incident of pollution, it is difficult to effectively implement the accountability, especially the local government and Party officials with decision-making power are hardly punished accordingly. As a result, the images of the Party and the government are damaged; the law loses its dignity, and the people lose their confidence. It is necessary to audit outgoing officials' natural resource asset management and establish a lifelong accountability system for ecological and environmental damage. "Any officials are not permitted to make a mess of the environment in some region and then simply leave the region, with still being officials and taking no responsibility."

3 Promoting Ecological Civilization Modernization of China's System and Capacity for Governance

Environmental problems are essentially related to the relationships between the overall situation and the local situation, and between short-term interests and long-term interests. These problems involve our economy, people's wellbeing, and political affairs. Since 1970s, we have proposed to avoid pursuing the old path of "developing the economy at the cost of the environment" prevailing in developed countries. This is our good wish. But now, it seems that we have not completely avoided it. This is mainly because:

First, from an international perspective, the pattern of economic development and ecological environment in the contemporary world runs through two fundamental problems in global relations—the contradiction between the East and the West, and the gap between the North and the South. Unbalanced economic development and ecological imbalance will continue. The main basis of this judgment is as follows: First, under the open market economy, poorer or relatively poor developing countries have no obvious comparative advantages in environmental technology, capital and human resources. From the view of comparative economics, these countries can only use their resources, environmental capacity and cheap labor to exchange and cooperate with developed countries on the basis of so-called "equality and mutual benefit". Second, the current international economic order is unreasonable. Specifically, resource consumption, pollutant emission, and use of cheap labor are all in developing countries, but the consumption of clean products is in developed countries. The sale price does not reflect the ecological cost accordingly. Therefore, the essence of today's international economic order is the cruel ecological plunder imposed by developed countries on developing countries.

Second, in China, the ecological problems in the east and west, cities and rural areas, suburbs and outer suburbs are increasingly taking on extremely complex governance patterns. Foreign media commented that the complexity of China's current situation is unmatched by any other country in the world. From China's east coast to the west inland, from prosperous cities to poor villages, from politics to economy, from society to culture, and from people's wellbeing to the environment, most of the phenomena occurring in western developed countries since the nineteenth century can be seen in China today. Due to China's unique development status and complexity, there is no country in the world whose successful experience can help China solve all its current problems. Because the challenges that China is facing now are as great as those that western developed countries have encountered in the past 200 years combined. The complexity and the particularity of historical missions are a tremendous pressure, a serious challenge, and also a powerful impetus to push us ahead. General Secretary Xi Jinping pointed out: "China's modernization is unique, unprecedented and extremely great. The total population of well-off countries in the world is less than 1.3 billion. If modernization is realized in China, with more than 1.3 billion people, the well-off population must more than double. It seems to be a dead end if we are to pursue the old path of consuming resources and pollute the environment! The reason why China's modernization is great lies in its difficulties we are facing. We cannot take the old path to catch up with well-off countries. The only path for us is scientific development." On the path of scientific development, what we can see clearly now is that as long as the development of any enterprise, industry or even region runs counter to the people's demand for improving the environment, its development path will inevitably become narrower and narrower, and eventually be "a dead end". Correctly understanding and dealing with environmental problems is not only a significant criterion for appraising whether officials at all levels are qualified, but also an important benchmark to evaluate citizens' moral integrity and cognition.

Third, in terms of the practice of building an ecological civilization, the difficulty for developing countries to solve environmental problems is beyond theoretical imagination, and is even more tremendous than that of developed countries to control environmental pollution. One reason is that we are facing the bottleneck in the development stage. At present, we are in the historical stage of rapid industrialization and urbanization. Objectively, the demand for resources and energy is high, and the trend of emission increase is strong. In this case, tasks of shifting the growth model and improving the economic structure are quite arduous and cannot be finished overnight. Between the late 1960s and early 1970s, when the United States declared war on environmental pollution, its urbanization rate had exceeded 75%, and the proportion of tertiary industries in the primary, secondary, and tertiary industries had reached 63%. The differences in development stages make it more difficult for us to solve environmental problems than developed countries. The other reason is that we are facing the bottleneck in natural endowments. The dialectical relationship between survival and development shows that within a certain limit, development is the perfection and promotion of survival; but beyond this limit, development will in turn pose a threat to survival. Development has a critical point, and approaching or exceeding the critical point will endanger the survival of mankind itself. This critical point mapped in nature, which includes both the scale and the speed of development, is the ability of the earth's ecosystems to get rid of the stale and take in the fresh and restore itself, or, the ecological threshold. General Secretary Xi Jinping profoundly pointed out: "The damage to the ecological environment mainly comes from the overexploitation and inefficient and blind use of resources. If someone drains the pond to get the fish in it, undoubtedly, there will be no fish left. Therefore, we must give top priority to resource use." "Conserving resources across the board is a fundamental policy to protect the ecological environment. It is better to attack a problem at its source than to apply a palliative, and this view should be especially established in protecting the ecological environment."

Based on this, we must follow basic strategies determined by the guiding principles of the 19th CPC National Congress and Xi Jinping Thought on Socialism with Chinese Characteristics for a New Era, and pursue a new path of promoting ecological civilization modernization of China's system and capacity for governance. General Secretary Xi Jinping pointed out: "The overall goal of deepening reform in every field is to improve and develop the system of socialism with Chinese characteristics and modernize China's system and capacity for governance."

A country's system and capacity for governance are the embodiment of the country's system and its enforcement capability. The national governance system is an institutional system that manages a country under the leadership of its ruling party, including institutions, mechanisms, laws, and regulations covering economy, politics, culture, society, ecological civilization and party building. That is, a set of closely linked and coordinated national institutions. The national governance capacity is the ability to use the national system to manage all aspects of social affairs, including reform, development, stability, internal affairs, foreign affairs, national defense, and governance over the party, the country and the military. The national governance system and capacity are an organic whole, which complement each other. Only with a good national governance system can the governance capacity be improved, and thus the efficiency of the national governance system be fully exerted.

To promote ecological civilization modernization of China's system and capacity for governance, we should focus on the overall situation, maintain an overall balance, identify priorities, consider both long- and near-term needs, and take comprehensive measures.

First, accelerating the strategic transformation of environmental management. This is the focus of modernizing China's ecological and environmental governance system and capacity. We will formulate and implement policies and measures based on the goal of improving environmental quality, and coordinate the relationships among pollution control, total emission reduction, environmental risk prevention, and environmental quality improvement so as to form a joint driving mechanism in which the improvement of environmental quality forces total emission reduction and pollution control, and then pushes the transformation of the growth model and improvement of the economic structure. We will constantly innovate the way of environmental management, change the focus from constraints to both constraints and incentives, and make more use of market mechanisms and means to guide enterprises' environmental behavior. We will promote pluralistic governance, improve social supervision mechanisms, strengthen the disclosure of environmental information, promote the sound development of social organizations of environmental protection, and establish an action system involving all sectors of society. We will apply more high-tech means to environmental protection, and make environmental management smarter and more detailed.

Second, deepening reform to push the transformation of functions. We will continue to reform governmental review and approval systems of environmental protection, pay close attention to canceling and delegating review and approval items that have been discussed and determined, and strictly regulate and control new governmental review and approval items. We will combine the reform of governmental review and approval systems with the procurement of services from social forces, broaden the supply channels of government environmental public services, improve the government's public service capacity, and promote the development of environmental protection industries, especially the environmental service industry. We will press ahead with government procurement of services in the fields of environmental impact assessment acceptance of projects, monitoring of environment quality, monitoring of discharge of pollution sources, environmental quality improvement and management technology, and sewage and household waste collection and treatment. We will explore new development paths of the environmental service industry, promote some provinces to carry out pilot projects of the environmental service industry, and guide local governments to open up the development of the environmental service industry.

Third, improving management institutions and mechanisms. Based on the systematic concept of restoring and maintaining the integrity and sustainability of the ecosystems, we will establish and improve an integrated ecological and environmental management system with the organic unity of functions, and coordinated and efficient operation, and realize the institutions, standards, and procedures of national ecological and environmental governance. We will establish and improve strict mechanisms for pollution prevention and control supervision, ecological protection supervision, nuclear and radiation safety supervision, environmental impact assessment, environmental law enforcement, and environmental monitoring and early warning. With breakthroughs in mechanisms, we will establish an environmental protection management system that uniformly supervises all pollutant emissions, and independently conduct environmental supervision and government enforcement of laws.

Fourth, enhancing environmental protection capacity. We will focus on building an advanced environmental monitoring and early warning system, a complete environmental law enforcement and supervision system, and an efficient environmental information support system to further improve the performance of environmental protection departments. We will improve the national environmental monitoring network, and enhance the coverage rate of environmental monitoring in rural areas. We will explore and promote regional joint law enforcement, strengthen the linkage between environmental protection and law enforcement by public security and courts, seriously investigate and deal with environmental violations, and enhance the deterrent effect of environmental law enforcement. Also, we will rely on IT application in national ecological and environmental protection, and promote disclosure of environmental information and resource sharing.

Chapter 8
Making Scientific Planning and Implementing the Blueprint Thoroughly

Since the 18th National Congress of the Communist Party of China in 2012, the CPC Central Committee with Comrade Xi Jinping at its core has deeply realized the great significance of urbanization to economic and social development. They scientifically planned and defined the development path, main targets and strategic tasks of urbanization in the future, and made a series of discourses and major policies and plans on deepening the development of new urbanization, forming a programmatic framework to guide the sound development of China's urbanization and move towards an urban society, and pointing out the direction for China's urbanization in the 13th Five-Year Plan period and even in the long term. The series of discourses and major policies and plans are the basic compliance for the development of new urbanization in accordance with the new requirements of "pursuing the path of new urbanization with Chinese characteristics and improving the quality of urbanization in an all-round way". This embodies the creative evolution and development of fine traditional Chinese culture and the modernization concept of building an ecological civilization, and is of great practical and far-reaching historical significance to completing the building of a moderately prosperous society in all respects and accelerating socialist modernization.

1 Following the Law of Urbanization and Scientifically Planning the Blueprint for Urban Development

Urbanization is the only way to modernization, and is the greatest potential of domestic demand and growth momentum in China. In the first year of the 13th Five-Year Plan period, General Secretary Xi Jinping made important instructions on deepening the development of new urbanization, emphasizing that the development of new urbanization must be based on a new starting point and make new progress. We should adhere to the development philosophy of innovative, coordinated, green, open and shared development, focus on urbanization of people, and pay more attention to

J. Pan, *China's Global Vision for Ecological Civilization*,
https://doi.org/10.1007/978-981-16-4534-1_8

increasing the whole registered population living in urban areas, ensuring equitable access to basic public services in urban and rural areas, maintaining a city's livable environment and historical features, and enhancing people's sense of fulfillment and happiness. We should follow scientific laws, strengthen top-level design, promote relevant supporting reforms in a coordinated manner, and encourage local governments to fit local conditions, highlight characteristics, and make bold innovations. Also, we should actively guide social capital engagement, and promote the sustained and sound development of new urbanization with Chinese characteristics.

Scientific planning is an important measure and successful experience in governing the country, and planning and promoting the sustainable and sound development of China's urbanization. As an important part of comprehensive development planning and the national planning system, scientific planning also plays a leading and normative role in China's development of new urbanization.

In accordance with the Five-sphere Integrated Plan, the overall plan for building socialism with Chinese characteristics,[259] we should follow the law of development, and promote the development of new urbanization in an active, prudent, solid and orderly manner. This means establishing a unified and standardized space planning system and governance system on the basis of following the national and regional functional zoning systems, regulating the use of all territorial space, and implementing systems for environmental protection. We should give full play to the functions of scientific planning, such as seizing the trend and taking action, making the most of the favorable conditions while averting unfavorable ones, guiding with strategies and standards, and taking rigid control. Also, we should improve the system and mechanism to promote the green, circular and low-carbon development of new urbanization, and develop spatial configurations, industrial structures, and ways of work and life that help conserve resources and protect the environment.

Planning of urbanization is a time–space yardstick for measuring the urbanization level in China. General Secretary Xi Jinping pointed out: "Urban planning plays an important leading role in urban development. Planning is the first focus in inspecting a city. Scientific planning brings biggest benefits, while mistaken planning leads to biggest waste, and random planning is the biggest taboo." As a policy tool and strategic starting point for further promoting new urbanization, scientific planning plays an important role in guiding, standardizing and regulating national economic and social development, optimizing urban and rural configuration, improving urban functions, guiding urban development in an orderly manner, improving urban development and management level, and enhancing people's wellbeing. From the perspective of urban scientific research, urban planning is an important basis and means to guide, standardize and regulate urban development, and protect and manage urban spatial resources. It plays a significant role in guiding the orderly development of cities and improving urban construction and management.

From the perspective of improving urban governance, it is proposed in the outline of the Thirteenth Five-year Plan that we will make innovations in the concepts and methods used in urban planning; properly define the scale, growth boundaries, development intensity, and protected space of cities; and strengthen planning and regulation to achieve multilayered urban spaces, coordination among different spatial layers, a coherent overall style and look, and cultural continuity. With regard to making innovations in the concepts and methods used in urban planning, promoting the development of new urbanization and realizing "scientific planning brings biggest benefits", putting people first is the core; seeking truth from facts is the prerequisite, and avoiding the waste caused by "mistaken planning", "random planning" and "taking a wrong path" is a necessary condition for achieving "scientific planning brings biggest benefits". It is necessary to make a blueprint for the development of new urbanization in line with the requirements of the times, the wishes of the people and the law of development through scientific planning and pragmatic actions.

Scientifically planning the blueprint for urban development, and understanding, respecting and abiding by the principles of urban development are significant guiding ideologies that must be followed. In the process of developing new urbanization, the fact that China is still and will long remain in the primary stage of socialism is the largest national condition that we need to face in promoting sustainable development and pursuing the path of new urbanization, and the contradiction between the people's ever-growing needs for a better life and unbalanced and inadequate development is still the main contradiction that we need to deal with in an overall way in promoting people-centered urbanization. China is the largest developing country in the world. We need to cultivate the ecological awareness of "taking mountains, rivers, forests, farmlands, and lakes as an ecosystem", and push China's urbanization towards a higher stage of civilization.

With regard to understanding, respecting and abiding by the principles of urban development, giving full play to the important leading role of urban planning in China's urbanization, adapting to the new requirements of China's development of new urbanization, and creating and cultivating new drivers of China's new urbanization, we need to take the following actions. First, we need to adhere to economic laws and scientific development, understand and grasp the great logic of China's economic development at present and for some time to come, and put them as the driving force and result of economic development, so that China's new urbanization planning and development can adapt to and lead the new economic normal; the disconnection between planning and market is avoided, and urban development and economic development can complement and promote each other. Second, we need to adhere to the laws of nature and sustainable development, and take the matching of population urbanization and land urbanization as constraints, so that the city scale and the resources and environmental carrying capacity can adapt to each other. Also, we need to establish a monitoring and early warning mechanism for the resources and environmental carrying capacity, and implement differentiated policies in finance, investment, industry, land, population, environment and assessment for different main functional areas. Third, we need to adhere to social laws and inclusive development, and ensure the matching of urbanization level and public service capacity. Fourth, we need to build a space planning system. On the basis of following the national plan for development priority zones and the national plan on urbanization for the 2014–2020 period, the functional zoning system should be implemented unswervingly; red lines for the protection of the ecosystems should be drawn, and the development should be promoted strictly according to the positioning of main functional areas.

We have reached a new historical starting point in planning the blueprint for our new urbanization, with a clear path and strong impetus. Just as General Secretary Xi Jinping showed full confidence in China's future: "We are closer, more confident, and more capable than ever before of making the goal of national rejuvenation a reality." Facing new characteristics of the times and practical requirements, we need to focus on scientific planning, provide a blueprint for development, and promote the delineation of ecological red lines. We should integrate the building of an ecological civilization into the whole process of the national development of new urbanization, and unify our thoughts and actions into the new development philosophy of innovative, coordinated, green, open and shared development based on following the fundamental basis of planning and development. We need to lead China's urbanization transformation under the new normal, and provide strong impetus for achieving the two centenary goals and the Chinese Dream of national rejuvenation, so that our people can share the development achievements of new urbanization.

New urbanization with Chinese characteristics is people-centered urbanization. At the central urban work conference, General Secretary Xi Jinping put forward that urban work should be based on "adhering to the people-centered philosophy of development", and emphasized that "urban management and services are significant criteria for judging whether urban work is done well, whether ordinary people are satisfied, or whether life is convenient". China's urbanization is promoted under the background of large population, relative shortage of resources, fragile ecological environment, and unbalanced development between urban and rural areas, which determines that urbanization cannot be simply equated with urban development. We must base our work on this most important reality—the primary stage of socialism, focus on the people-centered urbanization, follow the law of urbanization development, reasonably guide population flow, take well-ordered steps to the work on granting permanent urban residency to people originally from rural areas, work toward making basic urban public services cover all permanent urban residents, and constantly enhance population quality. This will enable us to realize a series of changes from "rural areas" to "urban areas", including industrial structure, employment mode, living environment and social security. We should promote well-rounded human development and social fairness and justice in the process of urbanization, and enable all residents to share the achievements of modernization, which are the basic requirements of new urbanization with Chinese characteristics.

The cities are built by the people. General Secretary Xi Jinping pointed out that our people have an ardent love for life and their aspirations for a better life is our goal. To lead the people to a better life is our Party's abiding goal. To conform to the people's aspirations for a better life and adhere to the people-centered philosophy of development, we should take "the cities are built by the people and for the people" and "ensuring the people's equal rights to participate and develop" as the starting point and ultimate goal in formulating the new urbanization plan, focus on ensuring and improving living standards, win the battle against poverty, and develop various social undertakings, so that gains of the development of new urbanization benefit all our people in a fair way.

The cities are built by the people and for the people. All urban development is meaningful only if it starts from the needs of the general public, focuses on people's needs, and responds to people's expectations. To promote new urbanization with Chinese characteristics and make farmers become citizens, we must put people first. Putting people first means regarding people as the basis and the core. Putting people first means that we should regard the people's needs for a better life and free and all-round development as the ultimate goal of all our efforts to achieve economic and social development of urban and rural areas. In the economic and social development of urban and rural areas, we should promote the urbanization process, take well-ordered steps to the work on granting permanent urban residency to people originally from rural areas, accelerate the comprehensive economic and social changes based on industrialization, agricultural modernization and rural urbanization, and realize the urbanization and modernization of rural areas, and free and well-rounded human development.

The development of new urbanization is a major systematic project. To scientifically plan the blueprint for urban development, it is necessary to find out breakthroughs for new urbanization in China to break new ground in urban development. General Secretary Xi Jinping emphasized that "urbanization" involves a wide range of work, and it should be actively and steadily promoted. The more complicated the work, the more we should focus on key aspects, so that making a breakthrough in one aspect can affect the overall situation.

Making overall planning and coordination of the three major structures of urban space, scale and industries. We will take city clusters as the main form to coordinate the three major structures of urban space, scale and industries, which is conducive to improving the overall situation of scientific planning. General Secretary Xi Jinping pointed out that we need to scientifically lay out production space, living space and ecological space, and take solid steps to promote ecological and environmental protection, so that a good ecological environment is helpful to improve people's quality of life and show our country's good image. The planning, construction and development of city clusters is the focus and highlight of urbanization development. General Secretary Xi Jinping pointed out: "We must put the focus of China's urbanization on promoting the development of city clusters." To take city clusters as the main form and scientifically plan urban spatial layout, we should do the following: First of all, we should promote the rational division of labor, complementary functions and coordinated development of large, medium and small cities and small towns, and build an open and efficient innovation resource sharing network, so that cities gradually form a development pattern of horizontal dislocation development and vertical division of labor and cooperation; Secondly, we should optimize and upgrade the eastern city clusters, create a number of city clusters and regional central cities in the central and western regions, and promote the coordinated development of border central cities and port cities.

Promoting the coordinated advancement of new urbanization and the building of new countryside. There are natural organic links in resources, economy, culture, ecology and environment between urban and rural areas. The areas are the most basic space for population and industrial clusters. To speed up the integration of urban and rural areas, work in urban areas needs to be promoted in step with the work of "agriculture, rural areas and farmers". We should scientifically plan towns and villages, improve planning and management of townships and villages, build beautiful countryside, and press ahead with the building of a new urbanization system with Chinese characteristics. This system is characterized by dual forces, which are supported by and supplemented each other, of "urbanization" and "developing villages and towns". General Secretary Xi Jinping pointed out: "Accelerating the integration of urban and rural development is a strategic task proposed at the 18th CPC National Congress, and an inevitable requirement for the implementation of the four-pronged comprehensive strategy." In the 19th CPC National Congress report, "pursuing a rural vitalization strategy" is further put forward. It is mentioned that we need to put in place sound systems, mechanisms, and policies for promoting integrated urban–rural development, and combine the development of massive economy with the promotion of urbanization and the coordinated development of regional economy, and the acceleration of the modernization of agriculture and rural areas. We will promote coordination between urban and rural areas, and the mutual integration and development of urban and rural areas in plans for the layout, allocation of production factors, industrial development, public services and ecological protection. What we aim to achieve is a new type of relation between industries and agriculture and between urban and rural areas in which industries promote agriculture, urban areas support rural development, agriculture and industries benefit each other, and there is integrated urban and rural development. We will gradually realize the equalization of basic rights and interests of urban and rural residents, urban and rural public services, and urban and rural residents' income. We will also achieve fair allocation of urban and rural production factors, and integration of urban and rural industrial development.

Taking a holistic approach in the three links of planning, construction and management. Taking a holistic approach in planning, construction, and management is the inherent requirement of further scientifically planning the blueprint for urban development. First, we must give full consideration to various factors such as urban functions, cultural characteristics and construction management when formulating plans. Second, the planning must be transparent. We must establish a regular mechanism and institutional arrangement involving the participation of representatives of planning enterprises and public institutions, construction contractors, management parties and citizens. Third, we must strengthen the openness and legal compulsion of regulatory detailed planning. Fourth, we must give priority to safety in our urban planning, construction and management, strengthen supervision in safety and quality, and ensure safety in all aspects of urban work and all areas of urban construction and development.

Unleashing and exploiting the potential of three driving forces of reform, science and technology, and culture. Scientific planning of the blueprint for urban development should focus on unleashing and exploiting the potential of three driving forces of local reform, science and technology, and culture. Specifically, we will promote the reform of planning, construction, management and household registration, strengthen the strategic research on granting urban residency to more people with rural household registration, and take orderly steps in enabling eligible regular residents in urban areas to localize their residency. We will create a comprehensive urban management database "smart city" and "sponge city" integrating population, resources, environment and culture. Meanwhile, we will protect, inherit and carry forward the legacy left by our predecessors and enhance urban competitiveness, so as to create conditions for realizing urban functions and goals.

Coordinating the three major layouts of production, life and ecology. What we pursue is the harmonious coexistence between man and nature, and scientific and balanced urbanization. To promote urban development, we should better deal with the relationship between man and nature. General Secretary Xi Jinping pointed out that in a limited space, with more construction spaces, there will be fewer green spaces, the self-circulation ability of the natural system will decline, and the regional ecological environment and urban living environments will deteriorate. We should consider the impact of every detail of urban planning and development on nature, and resolutely curb damage to the natural system. To coordinate the layouts of production, life and ecology, we should adhere to the concept of ecological civilization. Specifically, we should make scientific planning with "ensuring that the space for production is used intensively and efficiently, that the living space is livable and proper in size, and that the ecological space is unspoiled and beautiful" as necessary conditions. Based on regional natural conditions and ecological constraints of resources and environment, we should learn from mature experience, scientifically demarcate development boundaries and keep the pace of development under control, and create a macro layout of scientific and balanced urbanization.

Coordinating the three main bodies of government, society and citizens. To build new urbanization, it is necessary to keep all social forces including the government, society and citizens motivated in participating in urban planning of regions, implement the development philosophy, create synergy to promote the transformation of urbanization development mode, cultivate and carry forward core socialist values, enhance citizens' moral integrity, and accumulate positive energy for urban development. We should create a good living environment in pollution control and ecological restoration, improve people's living standards in poverty alleviation and sharing promotion, and enhance the urban governance capacity and operation in improving the urban governance system. Also, we should encourage enterprises and citizens to participate in urban planning, construction and management, and ensure that cities are jointly governed and managed, jointly built and shared by all.

2 Making It Possible for People to Enjoy the Natural Landscape

With regard to the development concept of new urbanization, General Secretary Xi Jinping profoundly put forward that urban development should embody the concept of respecting nature, following its ways, and unifying man and nature. Relying on the existing unique scenery, we should integrate urban development into nature, so that all people can enjoy the natural landscape. This is the inheritance and development of China's fine traditional culture and planning concepts. This is the sublimation of the guiding principles of the top-level design and planning of new urbanization by the CPC Central Committee with Comrade Xi Jinping at its core. This is also the people's aspiration for a better life. There are both historical ties and decision-making rationality.

Urbanization is the basic symbol of modern civilization, and the humanistic level of the development of new urbanization is the vitality of building cities with better life, creating beautiful countryside, and accelerating the modernization of agriculture and rural areas. Hometown memory takes roots in Chinese culture. Keeping hometown memory in our minds is the call and respect for our traditional culture, and also the foresight and major decision of the Party and the government in response to the popular feelings of the world. From the perspective of historical development, hometown memory is the cultural accumulation and continuation of the Chinese nation with "five thousand years of uninterrupted civilization", and is the humanistic embodiment of the Chinese people's unique spiritual thinking, characteristics of living in groups, and regional living habits. From the perspective of humanistic development, hometown memory is internalized as a habit, a memory and a spiritual medium under the unique rural features, ethnic culture and regional cultural characteristics, which not only shapes the unique rural features, but also gives farmers a lot of food for thought. To prioritize the development of agriculture and rural areas, we need to meet new expectations of farmers for a better life on the basis of following the general requirements of building rural areas with thriving businesses, pleasant living environments, social etiquette and civility, effective governance, and prosperity. In the practice and creation of agriculture and rural work, we need to improve the rural public cultural service system, innovate economic policies on the development of the rural cultural sector, and encourage creative transformation and development of the rural cultural sector.

Keeping hometown memory in our minds is closely related to the protection of traditional buildings and the shape of towns and villages. Without beautiful countryside, building a beautiful China would be impossible to practice. Only when every county, township and village become beautiful can building a "beautiful China" be a reality. General Secretary Xi Jinping emphasized that in promoting the development of urban–rural integration, attention should be paid to preserving the original features of villages, carefully cutting down trees, not filling lakes and demolishing houses, and improving the living conditions of residents in the original village form as much as possible. As early as when working in Zhejiang Province, General Secretary Xi

Jinping earnestly warned that "it is a pity that some old cities have been demolished and rebuilt, and that some cultural relics and historical sites have disappeared". In 2002, General Secretary Xi Jinping emphasized, in his preface to the book entitled *FuZhouGuCuo*, that "protecting old buildings and cultural relics means preserving history, maintaining a city's historical features, and conserving intangible fine traditions of historical and cultural cities". "Nowadays, in urban development of many cities, multiple old buildings are destroyed, enormous buildings with exotic styles emerge, and cities gradually lose their unique features".

Our society is changing all the time, and industrialization, modernization and urbanization have become an irresistible world trend. All these expanding areas need to be filled by skyscrapers and migrant workers. This will inevitably make many people who have left their homes feel homesick after material satisfaction. These areas should not only have gorgeous appearance, spacious quarters and unique modern styles, but also be closely related to local customs, folk customs and spiritual needs. They are expected to be "spiritual homes" to alleviate residents' psychological fatigue and enhance their spiritual happiness. These areas should be handsome and luxurious, and provide sufficient space for public activities. This kind of space can unleash our abundant feelings, rich creativity and sense of belonging. This kind of town, which fully integrates feelings, soul and culture, is of course the spiritual space of "keeping hometown memory in our minds".

Therefore, to promote the sustainable and sound development of new urbanization of "keeping hometown memory in our minds" and build beautiful countryside, we should not only pay attention to the protection and promotion of fine traditional culture in the physical and material sense, but also integrate the concept of making people's life more comfortable and modern elements. We should promote the scientific, balanced and fair allocation of urban spatial resources and layouts through spatial planning and market mechanisms. More attention should be paid to the continuation of urban and rural historical features in the socio-economic sense. This can help to embody the social development philosophy of people orientation, respecting regional culture and attaching importance to historical inheritance, and provide solid humanistic support and harmonious social foundations for finishing the building of a moderately prosperous society in all respects, speeding up socialist modernization and realizing the Chinese Dream of national rejuvenation. These are reflected in every detail of building beautiful countryside and cities with better life, as well as promoting the coordinated advancement of new urbanization and the building of new countryside.

To scientifically plan and promote the building of people-centered "beautiful countryside" and new urbanization, we need to foster a view of people and nature that "people can see mountains and water". Judging from the "hard" requirements of the development of new urbanization, "keeping hometown memory in our minds" is the spiritual requirement of the building of beautiful countryside and new urbanization, and also an inevitable choice for coordinating the coordinated and synchronous development of urban and rural areas and improving our farmers' happiness and satisfaction. Building rural areas, with thriving businesses, pleasant living environments, social etiquette and civility, effective governance, and prosperity, is the only path to attaining the goal of building a beautiful China. This is mainly reflected in four aspects: First, the environment of the rural areas must be pleasant, with green mountains, clear waters, blue skies and pollution-free lands. Dealing with ecological problems is the most common appeal of the Chinese people. Second, the rural areas must be equipped with integrated infrastructure with complete public services of urban and rural areas. We should actively work for the wellbeing of farmers and effectively improve their life satisfaction. Third, the rural areas must be underpinned by industries to make up for the shortcomings of rural development, increase rural people's income and ensure that a moderately prosperous society is built in all respects. Fourth, there must be cultural heritage in rural areas. We should conform to the historical trend of urban–rural integration, pay attention to the good natural ecological quality of the countryside, and highlight the local characteristics and humanistic environment.

Landscape with Chinese urban and rural styles is the external display of social and cultural genes in specific areas. General Secretary Xi Jinping pointed out that history and culture are the soul of a city, and we must protect the historical and cultural heritage of the city like cherishing our lives. Profound fine traditions of Chinese culture are our foundation to stand firm in the agitation of world culture. These propositions profoundly clarify both the significance of developing and promoting Chinese history and culture for national rejuvenation, and the mission and responsibility of developing and promoting Chinese culture in the process of new urbanization in China. We will accelerate the building of green cities and integrate the concept of ecological civilization into urban development. Propositions put forward by General Secretary Xi Jinping profoundly illustrate the goal of the planning of new urbanization and the building of "beautiful countryside". These propositions include: "We should consider the impact of every detail of urban planning and development on nature, and resolutely curb damage to the natural system", "taking mountains, rivers, forests, farmlands, and lakes as an ecosystem", "the countryside cannot become deserted and left-behind, and be a hometown in memory", and "building a beautiful China and beautiful countryside is to benefit our people, and every cent is used where it is needed most".

At present, China has been transitioning from a phase of urban and rural development with rural society as the mainstay to a new era with urban society and population as the mainstay. With regard to "keeping hometown memory in our minds", promoting the development of new urbanization, and strengthening rural revitalization, we should act on the following principles. On the one hand, we should take the building of pleasant living environments as our ultimate goal, get right the connection among production space, living space and ecological space with systematic thinking, organically integrate natural landscape into urban development, and encourage environmental protection and ecological restoration with natural restoration as the focus. With following the philosophy of green, circular and low-carbon development, we should well plan regional, inter-city and intra-city transportation, energy, water supply and drainage, heating, sewage, garbage disposal and other infrastructure networks and urban lifelines, and encourage the economical and intensive use of land, water, energy and other production input resources so as to make lucid waters and lush mountains appear again in cities and promote the formation of a green and low-carbon way of production and life, as well as mode of urban development and management. On the other hand, we should not lose our "culture". On the contrary, we should attach great importance to culture, especially Chinese culture, and make it embody Chinese elements and cultural genes. Besides, we should learn other advanced cultures with an open mind, learn from other cultural characteristics, create strong cultural atmospheres for our cities, enhance our cities' own attraction and charm, and build our cities into beautiful homelands with harmonious coexistence between man and man, and between man and nature so as to promote the harmonious development of our society.

From the view of scientific planning, "keeping hometown memory in our minds" should conform to our planning, and red lines for the quality of the environment and the protection of the ecosystems. In the *National New-type Urbanization Plan (2014–2020)*, it is proposed that according to the natural, historical and cultural endowments of different regions, regional differences should be reflected, and morphological diversity should be advocated. We should build beautiful towns with unique historical and cultural memories, regional features and ethnic characteristics, and form a practical and unique urbanization development model. With the advancement of new urbanization, the concepts of people orientation, respecting nature, and attaching importance to historical inheritance and green and low-carbon development will be gradually integrated into the entire process of new urbanization planning, the building of beautiful countryside, and rural revitalization. This not only adds a new concept for building cities with better life and creating beautiful countryside, but also embodies new urbanization requirements of people orientation and "keeping hometown memory in our minds". Urbanization is not to expand the city scale rapidly, but to "strictly control the increment and revitalize the stock". It is necessary to encourage the intensive use of land for construction projects, and enforce the red lines for farmlands and other resources and ecological conservation. We cannot expand land use for construction projects without restraint. Instead, we should "moderately reduce industrial land use" so as to protect agricultural areas such as farmlands, gardens and vegetable fields, and form a balanced layout of production, life and ecology. To keep hometown memory in our minds, we should attach great importance to ecological security. We will establish a monitoring and early warning mechanism for resources and environmental carrying capacity, expand the proportion of green ecological areas such as forests, lakes and wetlands, and enhance water conservation capacity and environmental capacity, so that people can enjoy the natural landscape in their daily life.

To keep hometown memory in our minds, we should think much of natural landscape, historical remains and cultural symbols when formulating plans. Based on this, we should know well the context of urban development and make it embody the concept of respecting nature, following its ways, and unifying man and nature. At present, in the process of China's active promotion of new urbanization, there are many problems, such as going in for grandiose projects, copying the experience of others, blindly relying on materialism, aiming high and going all out, sticking to individual ideas, and even pushing down the existing cultural heritage while transforming and building a cultural desert, which damages the natural landscape and cultural features of a city. Also, the large-scale demolition and construction in some rural areas also lead to the loss of local characteristics and folk culture. If a city makes the people who grew up here feel strange, it must be a huge loss of humanistic connotation to reposition their life. Therefore, in the process of accelerating urbanization, the decision-making should be democratic, the planning should be scientific, and the implementation should be prudent. Never concentrate on trivial matters to the neglect of important ones.

To keep hometown memory in our minds, we should work harder to protect cultural relics during the building of beautiful countryside and the promotion of rural revitalization. It is necessary to protect those historic buildings, traditional dwellings, ancient villages, famous trees, and even those exquisite street planning and street furniture. These are the ethnic characteristics formed by the Chinese nation for a long time, the customs to follow for generations, and the cultural memories passed down continuously for hundreds of years. From the view of practice, taking priority to planning and promoting it through projects are of crucial significance in building beautiful countryside. General Secretary Xi Jinping pointed out that the key to promoting the improvement of rural living environments is to take priority to planning. This demands that we should give full play to the normative and guiding role of planning in practice, and take the implementation of planning as the basic link of our work all the time, so as to achieve "compliance with laws, no overlap of coordinated efforts, and long-term use without waste". As regard to formulating plans for building beautiful countryside, we should give full consideration to local conditions, respect farmers' wishes, and pay attention to the flexibility of planning and projects. In the overall planning and layout, we need to combine the geographical location, resource endowment and industrial development of each village, consider the complementarity of business functions and undertaking of different tourist markets in space, and demonstrate and determine the key villages qualifying for being one of beautiful countryside, and core projects. We also need to optimize schemes, select the projects that are up to standard, carry out them in an orderly manner, and concentrate financial resources without distributing them unevenly. We need to attach importance to the feasibility of project implementation, avoid pursuing large and all-inclusive projects as well as large-scale demolition and construction, and encourage the intensive use of land.

To keep hometown memory in our minds, that is, in the development of new urbanization, we should be responsible for our history and people, adhere to the principles of going from past to the future and continuing ethnic culture on an ongoing basis, protect and inherit fine traditional culture, and retain unique features and historical heritage. On the one hand, we should, on the basis of following our national plans for development priority zones and important ecological areas, strengthen the planning and regulation of urban spaces with meeting the local requirements for historical inheritance, regional culture and the times. We should also integrate urban cultural continuity into multilayered urban spaces, coordination among different spatial layers, and the coherent overall style and look. It is imperative to protect the cultural heritage left by our predecessors, and retain the "genes" of our cities' unique regional environments, cultural characteristics and architectural styles, so that cultural relics and historical wisdom can be known to all of us. Besides, we should foster the natural features and spirit of the times of our cities and towns in a certain region and make united efforts so as to build a sense of pride and belonging and strengthen the confidence of the residents in this region. On the other hand, we should handle the relationship between urban transformation and development, and the protection and utilization of historical and cultural heritage, ensure that protection and development are inseparable, and delve deeper into and better elucidate China's excellent traditional culture. We should also carry forward the spirit of the times and strive to realize the creative evolution and development of Chinese traditional virtues, so as to stimulate the historical and humanistic heritage of a city and create a "golden name card" featuring a pleasant area for living, working and starting businesses for our cities.

3 Implementing the Blueprint Thoroughly

We will promote China's new urbanization to a higher stage and press ahead with the coordinated advancement of urbanization and the building of new countryside. On the one hand, as for scientific planning, we will manage overall and long-term development on the basis of establishing a system for territorial space development and protection, a space governance system, and a space planning system. Also, we will integrate multiple regulations into one, and ensure that one blueprint is implemented thoroughly. On the other hand, we will make laws and regulations serve as a "high-voltage line" to enhance the implementation and execution of the planning and development of China's new urbanization, and promote the transformation of the strategic guiding force, standard normative force and rigid control force contained in the blueprint of new urbanization into the realistic productivity that drives the urban development and transformation.

Planning is the soul and forerunner to lead the development of cities with better life and the building of "beautiful countryside". It is also the strongest productivity of cities and villages. Throughout the history of urbanization in the world, we can deeply understand that great planning and design can promote the function perfection and overall efficiency improvement of cities, towns and villages, and achieve sustainable development, but terrible planning and design will bring double losses of economic and social benefits.

Urban planning plays a role of strategic guidance and rigid control in the development of cities, towns and villages. A good planning helps to improve the efficiency of cities, towns and villages. To improve the functions of cities, towns and villages and achieve sustainable development, we need to strengthen the planning of cities, towns and villages on the basis of establishing unified and standardized space planning, improving functional zoning systems, and perfecting the system of regulating the use of all territorial space. We also need to enhance preemption, seriousness and continuity of the planning, so as to make the blueprint thoroughly, and more importantly, to implement a good plan to the end.

Cities, towns and villages are not only the basic space for people today, but also the production, living and ecological space for future generations. Therefore, scientific planning requires the thinking of "a long-term management" and the persistence of implementing the blueprint thoroughly. Only in this way can we build affordable, green, beautiful and livable cities and beautiful countryside, and can we inherit the spiritual features of cities and villages. To achieve this, we must first be responsible to the people and the future.

In a very long period of time, making frequent or unpredictable changes in planning is common. After frequent changes of planning, it is inevitable that the stability and continuity of urban development will be greatly reduced, resulting in serious waste of resources and funds. It is found that according to the ratio of demolition and construction of buildings in China during the 11th Five-Year Plan period, the annual demolition area of buildings in China during the 12th Five-Year Plan period is about 460 million square meters. If the service life of these demolished houses is reduced from 50 years of design to 25 years of actual use, based on an average of 1,000 yuan per square meter, China will waste 460 billion yuan per year due to premature demolition of buildings during the 12th Five-Year Plan period.

The chaotic planning stems from the concept of development. In some places, instead of seeking out local characteristics and endowments in a realistic way, local officials pursue large and strange architecture and believe that everything foreign is good. They rely entirely on their personal preferences and "intelligence" to guide local development. In addition, influenced by the achievement view of "grandiosity", some local leaders are bent on saving their own "landmark projects". In doing so, they thoughtlessly expand new areas for large-scale demolition and construction, and even cater to developers blindly in order to attract investment and win projects, ignoring local people's life. All of these are the embodiment of not making scientific planning according to urban planning, which makes the regional development lose its soul.

Undoubtedly, the difficulty in implementing the blueprint thoroughly is also closely related to the distorted view of political achievements of government and Party officials. In some places, government and Party leaders even make decisions alone in urban planning and development, and frequently say "I have the final say", replacing the rule of law with administrative orders. The old plans are reinvented and a batch of new plans are worked out, which is used by many newly-appointed local leaders to reflect their political achievements. They even call it "a new look, a new atmosphere". In fact, whether there is a new look or a new atmosphere does not depend on whether a new plan has been formulated. Combining with the new reality, using new ideas and new measures, turning the established scientific goals and good plans into reality in a down-to-earth manner is also the embodiment of a new look and a new atmosphere. Therefore, local government and Party leaders must further correct the view of achievement, carry forward the spirit of perseverance and truly implement the planning blueprint thoroughly so as to get somewhere.

General Secretary Xi Jinping pointed out that we have already got in our hands a good blueprint. What we should do is to follow it through to the end and make it a success. In this regard, we need to have a "nail" spirit. "When we use a hammer to drive in a nail, a single knock often may not be enough; we must keep knocking until it is well in place. Then we can proceed to knock the next one, and continue driving in nails till the job is completely done. If we knock here and there without focusing on the nail, we may end up squandering our efforts altogether." There is no use in saying, "I won't get the credit for success." If a blueprint is good, factually based, scientifically sound and well-received by the people we should keep working on it, one administration after another, and the outcome of our work will be real and appreciated and remembered by the people. Our officials should have a clear understanding of job performance, thinking more about working to lay a solid foundation which is conducive to long-term development and less about competing pointlessly with others, still less about building wasteful, showcase projects to prop up their own image. Let our officials be true and practical, dedicated to work and bold to shoulder their responsibilities, so as to live up to the expectations of history and the people.

To give full play to the government's regulatory function of scientific planning, we should adhere to the principle of planning before construction, grasp the orientation of regional development, and start construction only after every inch of land is clearly planned. First, we will delineate a large spatial pattern between regions on the basis of the overall strategy of regional development, pay attention to the management of development intensity, and integrate urban development boundaries and ecological red lines. Second, at the level of cities, towns and villages, we will upgrade their forms to enjoy more open space, making them reflect the characteristics of green, low carbon, intelligence, livability and suitableness for business. Third, we will innovate systems, mechanisms and policies and formulate supporting policies. Any success depends upon previous preparations. The importance of planning well and making sufficient plans is self-evident. In the construction of new urban areas, concerning urban orientation, land use, urban spatial pattern or urban form, it needs comprehensive consideration and careful planning. Any measures or strategies related to survival and development will not be accomplished overnight. Taking the integrated development of the Beijing-Tianjin-Hebei region as an example. General Secretary Xi Jinping emphasized that it is a major national strategy to realize the integrated development of the Beijing-Tianjin-Hebei region. It is a historic project and a realistic task facing China at present to plan and build Beijing Municipal Administrative Center, relieve Beijing of functions nonessential to its role as the capital, and promote the coordinated development of the Beijing-Tianjin-Hebei region. Only by strengthening faith, hard work, doing one thing at a time, working one after another, carrying forward the spirit of "craftsman", advancing our work carefully and in order, and leaving no historical regrets can we truly realize the vision of "better city, better life".

We should realize that planning must be scientific, because regional development is a comprehensive and systematic project, and the formulation of planning should conform to its development law. As an important policy to lead regional development, planning must be implemented rigidly.

To work for and manage the new urbanization, we must attach importance to planning and make it what really matters to all. As a public policy on safeguarding public interests, in the process of formulating and executing, more people's livelihood should be taken into account but not be hampered by the interests of the few; we must take a long-term view and show great foresight instead of changing our mind capriciously and doing project which tires the people and squanders resources; we must focus on the present and the future rather than sacrificing long-term interests for immediate ones. At present, the State Council has issued the document to strengthen the management of urban planning and construction and strictly claim accountability for violations of planning, which is to institutionally prevent inconsistent instructions and policy.

Planning is protected by law. "Urban and Rural Planning Law of the People's Republic of China" has been implemented since 2008. In the final analysis/at bottom, "top-level design" needs to be completed well in order to change the shortcomings of previous urban planning and stay the course. Urban planning should be formulated in accordance with the law, strengthen the examination and approval management of planning preparation in accordance with the law, strictly implement the principles and procedures stipulated in the Urban and Rural Planning Law, and conscientiously implement relevant provisions of the overall planning prepared by the level of government, participated by the public, deliberated by the Standing Committee of the National People's Congress at the same level, and examined and approved by the government at a high level. At the same time, urban planning approved according to law must be strictly implemented. Planning is mandatory, and any violation of planning must be seriously investigated for the responsibility of the relevant responsible person; The revision of the overall urban planning must also be approved by the original examination and approval authority, and reported to the Standing Committee of the National People's Congress at the same level for deliberation and approval, so as to prevent the random revision of the planning from the system.

The bottom line of urban planning is to ensure social stability, and the highest level is to guarantee the interests of all parties. "The continuity of urban planning shall be maintained, and the government shall not change as soon as it changes." "The opinions of the masses shall be listened to and the expert advice shall be respected in the preparation of spatial planning and urban planning that shall be determined through legislation after its formation, so that it has legal authority".[284] In the urban planning, we should give equal legal status and adequate expression channels to all rights entities, and promote the coordination and balance of multiple interests through scientific and democratic decision-making procedures. Therefore, it is necessary to rely on the rule of law to ensure that the blueprint is implemented thoroughly.

Actively promote urbanization planning legislation. Urbanization planning shall be a local legislative act based on development vision, scientific conception, rational layout and correct decision-making. Determining urbanization planning in the form of law is an important prerequisite for ensuring the stability and seriousness of urban planning, layout and administrative divisions, a basic basis for realizing efficient and rational allocation of resources, and a solid foundation for taking the road of people-oriented and sustainable development. Once the planning is formed and adopted, it shall be given legal status immediately, and no person or organization can change it at will.

Hold the red line of resources and ecology by law. New urbanization shall be based on resources conservation and environment protection. For a long time, in order to develop the economy, most local governments lack the bottom line awareness of ecological protection, environmental quality, resource utilization, and sacrifice the environment for urbanization. Therefore, we must gradually implement the strictest system of source protection, compensation for damage and accountability. Resolutely guard the lush mountains and lucid waters, and the red line of ecological protection, so that residents can still "enjoy the natural landscape" while enjoying the new urbanization achievements.

Establish a governance pattern of "Unity of four forces". New urbanization shall establish a reasonable and sound urbanization pattern in the form of law and clarify the relationship and role of government, enterprises and the public. Urbanization planning shall conform to the principle of openness and transparency. While considering the needs of local industrial development, the government shall fully respect the opinions of enterprises and the public and listen to the advice of experts in various fields, and implement publicity and hearing system for urbanization planning and adjustment, so that the public has enough right to know, participate and supervise the planning. In the implementation process of planning, we should give full play to the rule of market in resources allocation on the premise of keeping the red line of resources and ecology, and finally realize the governance pattern of "Unity of four forces" of government leading force, enterprise main force, market allocation force and social synergy competency in new urbanization.

Actively promote "multi-planning coordination". Local government shall actively promote the "multi-regulation coordination" of economic and social development, land use planning, urban and rural development planning, ecological protection planning, etc., and rationalize the relationship between departments, so as to make local economic and social development planning, urban overall planning and land use planning carried out on a common planning platform. Ensure the organic connection between the recent urban planning and the overall urban planning to realize the consistent time sequence. Strengthen the organic connection between the special urban planning and the overall urban planning to achieve the consistent target. Implement the organic connection between detailed planning and overall urban planning to realize the consistent statistical indicators. All kinds of planning are connected with each other and support each other, forming a planning system for urbanization construction.

Establish the system of environmental protection responsibility investigation for officials and natural resources auditing for officials when leaving office. During the process of urbanization planning and construction, environmental protection responsibility investigation system shall be established according to law. Officials who make blind decisions regardless of the ecological environment and cause serious consequences shall be held lifelong liability. Officials at all levels shall be guided to establish a correct view of performance. At the same time, the training of environmental protection legal knowledge shall be strengthened to enhance the sense of responsibility for ecological environmental protection of officials at all levels. Establish a balance sheet of urban natural assets and liabilities, and clarify the responsibilities of local leaders for ecological environment protection, cultivated land protection and land requisition, mineral resources development, and implementation of the system of paid use for natural resources, etc. Incorporating the implementation of urban planning into the assessment and outgoing audit of local party and government officials must be strengthened in the construction of new urbanization.

Chapter 9
Building a Green Home and Persisting in Promoting the Building of a Community with a Shared Future for Mankind

There is only one Earth for mankind, and building a green home is a common dream of the whole world. General Secretary Xi Jinping has always focused on China's position, world vision and human mind, and promoted the building of a community with a shared future for mankind. He explicitly pointed out: "The construction of ecological civilization id related to the future of mankind, and the international community shall joint hands to seek the road of building a global ecological civilization"[285] "We must accelerate the building of an ecological civilization from a global perspective, and turn green development into new China's overall national strength and new advantages in international competition."[286] While China is committed to the domestic ecological conservation, it maintains global ecological safety with the image of a major and responsible country, interprets the concept of global sustainable development with advanced ideas and positive actions, gradually becoming an important participant, contributor and leader in building global ecological civilization. This is the full display of the CPC Central Committee with Comrade Xi Jinping as its core in promoting the building of global ecological civilization, practicing the concept of green development and enhancing the capacity for modern governance in the international area.

1 A Community with a Shared Future Inter-Connected with Each Other

With the deep development of world multipolarization, economic globalization, cultural diversity and social informationization, countries around the world are closely linked. At the beginning of his tenure as president of the country, General Secretary Xi Jinping put forward the idea of "firmly establishing a sense of community with a shared future", emphasizing that common development is an important foundation for sustainable development, which is in line with the long-term interests and fundamental interests of people of all countries,[287] and believes that "While the

J. Pan, *China's Global Vision for Ecological Civilization*,
https://doi.org/10.1007/978-981-16-4534-1_9

world we inhabit is characterized by diversity in terms of culture, ethnicity, skin colors, religions and social systems, people across the world increasingly find themselves in a community with a share future with their interests interlocked, similarities and differences between countries coexisting, and interests and conflicts intertwining".[288] Once the initiative of the community with a shared future for mankind was put forward, it was generally recognized and welcomed by the international community, and it was written into many United Nations documents and became an international consensus.[289] Guided by the concept of "a community of shared future", China has created an all-round diplomatic layout, actively promoted the reform of global governance system, and has always made unremitting efforts to safeguard world peace, contribute to global development and maintain international order.

The report of the 18th National Congress of the CPC clearly put forward the idea of "advocating the consciousness of a community with a shared future for mankind" in dealing with international relations, deepening the diplomatic strategy of mutual benefit and win–win cooperation. The report of the 19th National Congress of the CPC clearly regards "building a community with a shared future for mankind" as the thought and basic strategy of socialism with Chinese characteristic for a new era, and put forward the overall objectives of "building a community with a share future for mankind and building an open, inclusive, clean, and beautiful world that enjoys lasting peace, common security, and shared prosperity", and pointed out that "We must keep in mind both our internal and international imperatives, stay on the path of peaceful development, and continue to pursue a mutually beneficial strategy of opening up. We must uphold justice while pursuing shared interests, and will foster new thinking on common, comprehensive, cooperative, and sustainable security. We will pursue open, innovative, and inclusive development that benefits everyone; boost cross-cultural exchanges characterized by harmony within diversity, inclusiveness, and mutual learning; and cultivate ecosystems based on respect for nature and green development". The general layout and general path of building a community with a shared future for mankind are comprehensively set forth from five aspects: establishing partnership, creating a security pattern, seeking development prospects, promoting civilization exchanges and building an ecological system.[290]

The "community with a shared future" is an international view that transcends nation-state and ideology.[291] Based on the dialectical philosophy of universal connection, China regards the world as a whole formed by the interconnection between countries. All countries are closely related to each other, sharing life and death. We should focused on the interests of all mankind, adhere to the unity of discussing global issues and achieving common prosperity, respect the differences between different cultural backgrounds and different development paths, seek common ground while reserving differences, so as to build a fair and rational new international order. The "community with a shared future" is also a mutually beneficial and win–win development concept, which originates from China's traditional thought of "harmony in diversity". It not only advocates abandoning zero-sum games thinking, pursuing cementing good neighbourliness and friendship and realizing cooperation and win–win, but also emphasizes making a difference, "making the world better", jointly coping with challenges and sharing development opportunities, creating a peaceful and stable international environment and establishing a "new international relationship" with "cooperation and win–win" as the core. The "community with a shared future" includes five major pillars: politically establishing a partnership of equal treatment with each other and mutual consultation and mutual accommodation; creating a security pattern of fairness, justice, joint construction and sharing in terms of security; economically seeking open, innovative, inclusive and mutually beneficial development prospects; culturally promoting civilization exchange of "harmony in diversity" and inclusiveness; environmentally building an ecological system that respects nature and develops green.[292]

Extensive economic development mode under the guidance of industrial civilization is unsustainable. Both developed industrialized countries and developing countries that have not yet completed industrialization are aware of the need to abandon traditional development ideas. China's ecological civilization construction just provides systematic theory, method and policy experience, which is the world outlook, values and methodology of the whole human society in terms of the harmonious development between human and nature, society and nature. It will lead to the change of value cognition, production mode, consumption mode, and the corresponding institutional mechanism. It is the world's demand that sustains China's ecological civilization construction scheme of "the community with a shared future for global human". With the aim of "building a community with a shared future", China actively develop global partnership, expands the intersection of interests with other countries, and promotes the establishment of a global governance system characterized by joint consultation, joint contribution and shared benefits. The concept of a community with a shared future was first practiced in regional cooperation in Asia, and the initiative of "ASEAN-China community with a shared future" provided support for the connection between Southeast Asia and China; Building a "community with a shared future in its neighborhood" and deepening the relationship with its neighbors according to the concept featuring amity, sincerity, mutual benefit and inclusiveness and its neighborhood diplomacy of building an amicable relationship and partnership with its neighbors; The proposition of "Towards a community of common destiny and a new future for Asia"[293] maps out a brighter future for promoting the regional cooperation in Asia and jointly building a green home. China will continue to send a "green invitation letter" to the world,[294] by actively promoting the Belt and Road initiative on international cooperation and coping with climate change and other practical actions, so as to share opportunities and seek development with other countries in the world.

2 Integrating Green Development Concept in the Construction of the Belt and Road Initiative

The Belt and Road Initiative, a realistic portrayal of building "a community with a shared future for mankind", is put forward based on the realistic background of China's transformation of its development mode, boosting its development advantages, deepening international cooperation and broadening its development space, and conforming to the new trend of economic globalization and world scientific and technological revolution., reflecting the essential requirement of mutual benefit and win–win and opening a new chapter of intercommunication and mutual benefit between China and Europe, Asia and Africa. It not only covers the interconnectivity and intercommunication of policies, trade, culture and capital, but also integrates the ecological civilization into it, endowing the Belt and Road with a new meaning and making China share the concept of "green development" with countries along the route, to make developing countries get on board the express train of China's ecological civilization construction.

Integrating "green development" into economic, cultural and social construction along the route is a highlight of the Belt and Road Initiative. "Vision and Action on Jointly Building Silk Road Economic Belt and 21st Century Maritime Silk Road"[295] clarifies the overall framework and specific tasks of the Belt and Road initiative, and clearly integrates green development into various construction fields along the route: Highlighting ecological civilization concept, strengthening cooperation in ecological environment, biodiversity and climate change, and jointly building a green silk road in terms of investment and trade; Strengthening green and low-carbon construction and operation management, and fully considering the impact of climate change regarding to infrastructure construction; Promoting cooperation in clean and renewable energy such as hydropower, nuclear power, wind power and solar energy in the field of energy development; Enhancing in-depth cooperation in emerging industries such as technology, biology, new energy and new materials in industrial cooperation; Strictly protecting biodiversity and ecological environment in advancing local construction in countries along the route; Reinforcing exchange and cooperation among non-governmental organizations in countries along the route and widely carrying out various charitable activities such as biodiversity and ecological protection during non-governmental exchanges.

At the symposium on promoting the construction of the Belt and Road initiative, General Secretary Xi Jinping emphasized that we should focus on building the green Silk Road, the health Silk Road, the educational Silk Road and the peaceful Silk Road to make the construction of the Belt and Road initiative benefit the people of all countries along the route.[296]

The green Belt and Road initiative focuses on policy coordination, traffic infrastructure connectivity, unimpeded trade, financial integration and people-to-people bond, takes inclusive development as its essential feature, and sustainable development as fundamental goal. It aims to play the role of cooperation and exchange through international connectivity channels, strengthen ecological protection services and support for the construction of the Belt and Road initiative,[297] and promote countries along the route to change their development modes and jointly create new development opportunities. Its ecological connotation is mainly reflected in three aspects.

Firstly, we should spread the concept of ecological civilization and carry forward the ecological value of harmonious coexistence between human and nature. The Silk Road, which was established in the Western Han Dynasty, has been a channel for economic, trade and cultural exchanges between the East and the West since ancient times. The establishment of the Belt and Road inherits the traditional values of foreign exchanges and good neighborliness, practices the concept of justice and benefit of mutual respect and win–win cooperation, pays more attention to the dissemination of ecological civilization concepts in contemporary times, and advocates the harmonious coexistence between human and nature. Realizing sustainable development is the common goal of all countries around the world. Through spreading ecological civilization concept to countries long the route via the Belt and Road, China helps promote countries reach a consensus on ecological and environmental protection in cooperation and establish a coordinated development mode between economy and environment.

Secondly, we should promote the ecological economy to ease the dual pressure of economic development and ecological environment. The output of excess capacity and relatively low-end production capacity is not equal to polluted capacity, but should have a basic green threshold,[298] ensuring the coordination between development and protection through promoting ecological economy. The ecological economy characterized by "low energy consumption, less-polluting and high efficiency" requires the establishment of green industrial system and the promotion of optimization and upgrading of domestic industry, thus facilitating the "going global" of domestic green production capacity and driving the countries along the Belt and Road to achieve green production mode; it requires the clarification of green environment standard, specification of environmental impact assessment, and forestalling and defusion of environmental risks; it also requires to implement green credit based on the Silk Road Fund and Asian Infrastructure Investment Bank and provide financial support for environmental protection technology R&D, new energy development and utilization, emerging industry cultivation, green public goods and service construction.

Thirdly, we need to strengthen the eco-environment governance to ensure the safety of resources and energy and the stability of the ecosystem. Most of the countries along "the Silk Road Economic Belt" experience continental climate and alpine climate with dry climate and less precipitation, and have fragile ecological environment here owing to many plateaus and deserts and great population pressure. The "Maritime Silk Road" passes through many shipping hubs and energy transportation channels, and faces many threats to marine ecology. Countries along the Belt and the Road can learn from China's experience in environmental protection, strengthen joint research and development and exchanges on energy conservation and carbon dioxide emissions reduction, and technologies on environmental pollution control and on resource development and utilization, reduce the damage of economic activities to the eco-environment, and control the discharge of land-based pollutants into the sea from the source. We work together to build a defense for ecological security and carry out cross-border projects for eco-environment control, in order to maintain the stability of ecosystem functions.

Adhering to green thinking in the construction of "the Belt and Road" is of greatly practical significance to the development at home and abroad. Implementing the green "the Belt and Road" Initiative is the requirement of coordinating the overall course of at home and abroad, aiming at transforming green development into new overall national strength and international competitive advantages.[299] The "Belt and Road" connects the Asia–Pacific economic circle with the European economic circle, forming a large pattern of space development and regional cooperation, which promotes the circulation and allocation of manpower, financial and material resources in countries along the Belt and Road and adds vitality to regional development. The sustainable development mode followed by the green "Belt and Road" meets the transformation and upgrading development needs generally faced by Asian and European countries. "Green is related to the well-being and future of all mankind, and it also breeds historic opportunities for world development. "Green dream is not only Chinese, but also global."[300] China has fully implemented the concept of green development into the Belt and Road initiative and led the countries along the Belt and Road to common prosperity,[301] which is not only the demand of transforming the development mode, deepening reform and enhancing international competitiveness, but also the embodiment of China's extensive participation in international cooperation and establishing the image of a responsible big country, conforming to the overwhelming trend of peace, openness, inclusiveness, cooperation and win–win.

Implementing the green "Belt and Road" initiative is conducive to broadening the development space, creating the development opportunities and achieving mutual benefit and win–win results. Geopolitical "survival space theory" holds that a country is like an organism, and its prosperity needs the space to support its survival and development. At present, China's economic growth is going through a shift period from high-speed to medium-speed. The obvious overcapacity problem, increasingly tight constraints on energy resources and urgent task of energy conservation and carbon dioxide emissions reduction limit the space for economic development. Green "Belt and Road" promotes the integration and upgrading of industries such as iron and steel, building materials, railway and electric power, and transfers to countries needing such products and technology, which will effectively solve the problem of domestic overcapacity. Through transformation and cooperation of green economy and green technology, China works with countries along the Belt and Road to jointly cope with ecological and environmental risks, broadening the domestic development space, creating investment opportunities and development dividends, promoting the circulation of regional factor, upgrading of industrial structure and the governance of ecological environment, and achieving mutual benefit and win–win.

Implementing the green "Belt and Road" is the fundamental guarantee for pooling public mind and maintaining mutual benefit and interconnectivity. The interconnectivity process advocated by the Belt and Road is also a process of seeking consensus between China and countries along the Belt and Road. To maintain the long-term mutual benefit and interconnectivity, it is necessary to comprehensively consider the common economic, social and ecological interests of all countries, fully measure the ecological and environmental impacts during the implementation of major projects such as industrial layout, energy development, infrastructure construction and land planning, formulate response measures for environmental risks and improve production situation of countries along the Belt and Road and the living standards of residents in an all-round way. For example, we need to clearly define the standards of water, land, material saving and environmental protection in infrastructure construction, emphasize resource value and clarify property rights in resource development and utilization, support the implementation of energy-saving and environmental protection projects by green finance in financing. The green concept shall be expanded to all aspects of the Belt and Road construction, making the Belt and Road a regional cooperation development model with infrastructure construction as the carrier and sustainable development as the criterion.

With the green Belt and Road as the link, China will lead countries along the Belt and Road to carry out strategic and advantageous complementarity, explore the sustainable development path with equal emphasis on ecological and economic benefits, and release the potential of cooperation development by giving full play of comparative advantages of all countries, so that the Belt and Road construction will benefit the people of all people along the Belt and Road.

3 Guiding International Cooperation on Climate Change and Maintaining Global Ecological Security

Climate change is the overwhelming global environmental problem, and it is also the biggest challenge faced by environmental managers. To deal with climate change, the international society needs to work together to build a global climate governance system to resolve conflicts and jointly commit to global sustainable development. As the biggest developing country, the largest consumer, producer and net importer of energy resources,[302] and the largest carbon emitter,[303] China plays a significant role in global climate governance, guiding international cooperation on climate change and fully demonstrating responsibilities of a major country.

The essence of climate change is development, which is related to ecological and environmental, economic and political interests. It contains the value of fairness and justice and promises profound theoretical basis.

First of all, coping with climate change involves global morality. "Climate change" refers to the climate change caused by direct or indirect human activities changing the composition of the Earth's atmosphere in addition to the natural climate variation observed in similar periods. The core of climate change is global warming, which will cause changes in natural environment or biota, exerting a greatly harmful impact on the composition, resilience or productivity of ecosystem, the operation of socio-economic system and human health and welfare. Due to the complexity and irreversibility of its adverse effects, it is necessary for all countries to jointly hold the responsibility of governance and maintain a good living space for present and future generations.

Secondly, climate change negotiation is an international political game for development space. At present, under the energy consumption structure dominated by fossil fuels and limited technical conditions, economic activities will inevitably involve carbon dioxide emissions, and carbon dioxide emissions reduction requires financial and technical support, which increases development cost in the short term, especially for developing countries that have not yet realized industrialization. The task of realizing economic development and emission reduction at the same time is formidable. Therefore, the competition for carbon emission space among countries means the competition for development space, the fairness and responsibility should be fully considered and we must do the best we can.

Thirdly, coping with climate change provides development opportunities for all countries. The urgency of global climate governance requires that we must avoid the precedent of developing the economy at the cost of the environment. Carbon tax, carbon emissions trading rights and other carbon emission control measures increase production cost and thus affect the competitiveness of a country's enterprises: In international competition, whoever can master carbon emission reduction clean technology and adopt green recycling and low-carbon economic form means who will occupy the commanding heights of the world economy. In brief, the problem of climate change is essentially a development problem, and the process of implementing carbon dioxide emissions reduction is conducive to the transformation from extensive development mode to sustainable development mode, relying on technological innovation to develop new economic growth points, open up new markets and achieve a win–win situation for economy and environment.

Finally, coping with climate change needs all countries to share the responsibility of carbon dioxide emissions reduction and implement global governance. President Xi Jinping said that "Climate change is a global challenge, and no country can stay immune to such impacts",[304] pointing out the global and necessary significance of climate governance. According to the assessment of the Intergovernmental Panel on Climate Change (IPCC), the increasing atmospheric greenhouse gas concentration caused by human activities is the main factor leading to global warming. Therefore, the core measure to deal with climate change is to reduce carbon dioxide emissions.[305] However, carbon dioxide emissions reduction is the largest public product in the world,[306] which is non-exclusive and non-competitive. The lack of institutional constraints will inevitably drive countries to "take free riding" and they will be unwilling to take the initiative to undertake the responsibility of emissions reduction. The adverse effects of climate change caused by excessive emission of greenhouse gases must be paid by all countries in the world. Climate change poses severe threat to the development space of all countries. The complex and global significance of climate change and public good attribute of climate change determine that the international community should establish a global governance mechanism to ensure the implementation of emissions reduction tasks and jointly cope with climate change risks. The Commission on Global Governance defines "governance" as the sum of many ways in which individuals or groups handle their common affairs, which is a continuous process of reconciling conflicting or different interests and taking joint actions.[307] Therefore, global climate governance is embodied in pooling the strength and coordinating the interests of all countries, taking reducing greenhouse gas emissions as the core task, reaching consensus on emissions reduction time, standards, funds, technology and supervision, forming binding agreements and putting them into action, and gradually mitigating and adapting to climate change.

Although US President Trump announced his intention to leave Paris Agreement, which has hindered the global response to climate change to a certain extent, it is still a global consensus and strong political will to deal with climate change. In this regard, the Chinese government stated that no matter how other countries' climate-related policies change, China, as a responsible major country, will not change its determination, goals and policy actions to deal with climate change, and China is willing to work with relevant parties to jointly safeguard the achievements of the Paris Agreement to deal with global climate change. China takes active response to climate change as its due international obligation, and has shown sincerity, determination and Chinese wisdom in negotiations concerning climate change and actions related to climate governance. Moreover, it also provides experience accumulated in domestic eco-environmental construction and green transformation for global climate governance. China has achieved a balance between realizing its national development interests and safeguarding the interests of all mankind, and has made commendable contributions to the world on the international stage. To contribute its wisdom and promote inclusive development, China follows the principle of pursuing the greater good and shared interests and seeks the broadest possible common ground of all parties involved to promote international cooperation on climate governance. President Xi Jinping advances toward an equitable climate governance regime that is cooperative and beneficial to all, and advocates that countries "give full play to the strengths and promote win–win cooperation", "pursue the rule of law and fairness and justice" and "follows the principles of inclusiveness, mutual learning and common development"[308] in the global governance. Under this concept, countries can find the best response starting from their own national conditions,[309] which is in line with the traditional Chinese culture's ideas of "seeking harmony but not uniformity" and "when big rivers have water, the small ones are filled; and when small rivers have water, the big ones are filled". President Xi Jinping emphasized that climate governance is not a "zero-sum game", and coping with climate change is a common cause of mankind, during which developed countries take the initiative to reduce emissions, while developing countries should also avoid taking the old road of industrial civilization with high carbon emissions.[310] On the eve of the Paris Conference, China held negotiations with many countries and issued a joint statement. The differences in the focus of climate negotiations, such as legal binding force, funding and intensity, were described in the joint statement, which provided a political basis for the successful convening of the Paris Climate Conference.[311] In the conference, President Xi Jinping stated that the standard of a successful international agreements lies in solving current contradictions as well as guiding the future. Based on this judgment, President Xi Jinping pointed out that the Paris Agreement should both focus on strengthening the global action on climate change after 2020 and inject impetus into the promotion of a better achieving on a sustainable development all over the world. In particular, the Paris Agreement should be good for achieving the objective of the Convention and leading a green development; be good for gathering global strength and encouraging a broad participation; be good for increasing resource input and strengthening action guarantee; and be good for pursuing pragmatism and effectiveness starting from the national conditions of each country. The four proposals provide guidance for solving the practical problems involved in the Paris Agreement from the following aspects: basic conditions, institutional arrangements, financial and technical support and adherence to common but differentiated responsibilities.

Showing China's stance and fulfilling its emissions reduction commitments. President Xi Jinping pointed out that "The greatest determination will produce the highest wisdom", "All parties should show sincerity, strengthen confidence and make concerted efforts in climate change negotiations"[312] and "China has been actively responding to climate change in a responsible manner and is willing to continue to assume international responsibilities consistent with its prevailing national conditions, the stage of development at which it stands and actual capabilities".[313] China takes the initiative to undertake the responsibility of emissions reduction. In the *Intended Nationally Determined Contributions*, China has made a set of commitments: to achieve its peak carbon dioxide emissions around 2030 (and to make every possible effort to peak earlier); to lower carbon dioxide emissions per unit of GDP by 60–65% from the 2005 level; to increase the share of non-fossil fuels in primary energy consumption to around 20%. At the High-level Signature Ceremony for the Paris Agreement, China has unveiled its goal of investing 30 trillion yuan from 2016 to 2030 to cope with climate change in the Nationally Intended Nationally Determined Contributions. General Secretary Xi Jinping emphasized: "In the future, China will further strengthen the control of greenhouse gas emissions and strive to achieve the goal of reducing carbon intensity by 40–45% by 2020".[314] China and the United States are the world's two largest carbon emitters, and their climate governance actions are directly related to the effectiveness of global climate change negotiations. China and the United States have issued joint communiqués on climate change three times, and China first put forward the goal of peaking carbon emissions in 2030.[315] Both countries have made a commitment to give priority to funding by public resources and encourage the gradual adoption of low-carbon technologies, and proposed that they take steps to join the Paris Agreement as soon as possible.[316] Prior to the G20 Hangzhou Summit, China fulfilled its commitments and ratified the Paris Agreement, which effectively promoted the entry into force of the Agreement and set an example for other countries to fulfill their commitments.[317]

Implementing China's measures and helping developing countries reduce emissions. President Xi Jinping pointed out: "With different historical responsibilities for climate change, developed countries and developing countries have different development needs and capabilities. Just like a racing, some cars have already run a long way, and some cars have just set off. At this time, it is inappropriate and unfair to limit the speed with a uniform scale. Developed countries have set an example in addressing climate change, which is in line with the important principles of common but differentiated responsibilities, fairness and respective capabilities established by the United Nations Framework Convention on Climate Change, and is also the common aspiration of developing countries."[318] Upholding common but differentiated responsibilities has always been the standpoint for China to promote global climate governance. At the same time, "China will shoulder its share of responsibility and continue to play its part. We also urge developed countries to fulfill their historical responsibility, honor their emission reduction commitments and help developing countries mitigate and adapt to climate change".[319] China not only firmly defends the basic right to development of developing countries in the climate change conference, but also provides practical assistance to developing countries in terms of green technology transfer, financial support and education. At the United Nations Sustainable Development Summit, General Secretary Xi Jinping advocated that all countries strengthen cooperation, jointly implement the UN's 2030 Agenda on Sustainable Development, strive to achieve win–win cooperation, and announce practical assistance to developing countries, including establishing the South-South Cooperation Assistance Fund and continuing to increase investment in the least developed countries. For the least developed countries, landlocked developing countries and small island developing countries, the debt they have incurred in the form of interest-free Chinese government loans due to mature by the end of 2015 will be exempted. International development knowledge center will be set up to explore the construction of global energy internet to promote clean and green approaches to meet global power demands.[320]

Acting out China's responsibility and building a platform for international cooperation. China mediated at the international climate conference to promote the formation of agreement, reflect the interests of developing countries and help developing countries balance the pressure between emissions reduction and development. At the same time, we will spare no effort to fully grasp the opportunities of international exchanges and cooperation, and rely on the strength of intergovernmental organizations or non-governmental organizations to build new platform and form new mechanism. For example, taking Tianjin APEC High Level Roundtable on Green Development as a platform, the global green supply chain and value chain cooperation initiative was initiated and implemented,[321] which promoted the industrial upgrading and green development mode transformation, and provided ideas for the construction of global green industrial system. At the Second Session of the China-U.S. Climate-Smart/Low-Carbon Cities Summit, representatives of China-U.S. subnation, cities, research institutions and enterprises in China and the United States conducted in-depth exchanges and discussions on 17 topics including carbon market and city emission peaking, and signed cooperation agreements on low carbon city policy research and capacity building, and low carbon technology innovation and application[322]; At the same time, at the Third China-US Governors' Forum, General Secretary Xi Jinping pointed out that "In terms of environmental protection, China has the demand and the market, while the US has the technology and the expertise. Exchange and cooperation in environmental protection at the local level should be part and parcel of our joint efforts to address climate change and promote sustainable development", encouraging local efforts to promote exchanges and cooperation between the two countries and implement emission reduction actions.

Sharing Chinese experience with other countries and promoting ecological progress. On many international occasions, General Secretary Xi Jinping indicated that China will promote climate change as an important part of ecological civilization construction, and promote the advanced concept of green development to the international community: "China regards addressing climate change as a great opportunity to realize the transformation of development mode, and actively explores a low-carbon development path in line with China's national conditions. The Chinese government has fully integrated climate change into the overall strategy of national economic and social development".[323] "China will pay more attention to green development, integrate ecological civilization construction into all aspects and the whole process of economic and social development, and strive to achieve sustainable development and comprehensively improve its ability to adapt to climate change".[324] The report to the 19th CPC National Congress raised the construction of ecological civilization to the national strategic height, making it an important path to establish a sustainable development model. It clearly stated that the energy consumption and carbon dioxide emissions per unit of GDP as well as the discharge of major pollutants should decrease sharply, and advocated "actively coping with global climate change together with the international community".

4 Jointly Building a New Global Governance Order and Deepening Global Sustainable Development

Achieving sustainable development is the common goal of human society. From the concept of "sustainable development" put forward by the World Commission on Environment and Development for the first time in 1987, to the establishment of a series of sustainable development action plans such as Agenda 21, Millennium Development Goals and 2030 Agenda for Sustainable Development, the pace of global sustainable development has been kept moving forward, which has a far-reaching impact on the development strategies and development space of countries. In conforming to the trend of international sustainable development, China has continuously deepened its understanding of sustainable development, explored correct development ideas in light of particular national conditions, promoted economic transformation internally, strengthened the construction of ecological civilization, and implemented the 2030 sustainable development goal with positive actions; China has also established dialogue mechanism externally, promoting cooperation and exchanges, turned the initial participant to the leader and promoter of global governance from, made remarkable achievements, and achieved the integration of promoting the global sustainable development process and implementing China's ecological civilization construction.

The Agenda 21, which was passed at the United Nations Conference on Environment and Development in Rio in 1992, was the first global action plan for sustainable development. Due to its wide coverage, poor relevance, lack of quantification and failure to consider the national conditions of various countries, its implementation was ineffective. Based on this, China has formulated China's Agenda 21, which serves as the action program for China to implement the sustainable development strategy, and initiated the preliminary exploration of the sustainable development strategy.

In 2000, the United Nations Millennium Summit launched the Millennium Declaration, and eight specific Millennium Development Goals (MDGs) was raised the following year, which gathered global efforts to solve key problems such as poverty reduction. Each goal was quantified and set a deadline of 2015, which was greatly practical. In the year finishing MDGs, according to the Report of the Millennium Development Goals (2015), China has achieved the best implementation results and made the greatest contribution to the world. At present, it has basically completed the Millennium Development Goals and made remarkable achievements in poverty reduction, health, education and other fields, mainly including halving the proportion of people with per capita income less than \$1.25 from 1990 to 2015, and ensuring that all children complete primary education courses. The under-five mortality rate has been reduced by two thirds, the maternal mortality rate by three quarters, and the proportion of people without sustainable access to safe drinking water and basic sanitation facilities has been reduced by half.[325]

The 2030 Agenda for Sustainable Development adopted by the United Nations in 2015 further emphasizes that environment, economy and society are the pillars of sustainable development. At the same time, it also incorporates the means and content of implementation, paying attention to the relevance and unity of various fields and purposes, and adopting global indicators to measure and monitor with regional or national indicators drawn up by Member States. It is more universal and operable, leading the world to eliminate extreme poverty, overcome inequality and injustice in the next 15 years of development. The "5P" (People, Planet, Prosperity, Peace, Partnership) vision put forward in the Agenda embodies the integration thought of people-centered, protecting the earth, developing economy, social harmony and win–win cooperation. Its wish of "never let anyone fall behind" and "common but differentiated responsibilities" provides opportunities for developing countries to participate in global sustainable development governance in depth. China took the lead in publishing Position Paper of the People's Republic of China on the Implementation of the 2030 Agenda for Sustainable Development, which established the principle of "promoting the development of economy, society and environment in a coordinated manner, and realizing the harmonious coexistence between man and society and man and nature", and made arrangements in key areas such as eliminating poverty and hunger, strengthening environmental governance, promoting the protection and restoration of natural ecosystems, making every effort to cope with climate change and effectively utilizing energy resources. 2016 is the first year of the 13th Five-Year Plan. Strategic arrangements for the implementation of China's sustainable development agenda from both domestic and foreign aspects was made in the 13th Five-Year Plan outline, realizing the organic integration of the implementation of sustainable development goals with the national medium and long-term development plan [2]. China seized the opportunity provided by the 2030 Agenda for Sustainable Development, actively participated in international affairs, spread the concept of ecological civilization, and made greater contributions to promoting global sustainable development.

Although China has achieved certain phased results in the process of sustainable development, there are still uncoordinated problems in the development, especially in the current shift period of economic growth, where resource constraints are tightening and people's livelihood needs to be improved, and all three are mutually constrained. Therefore, it is necessary to seek new impetus and new mode to break through the bottleneck of development. Under this condition, General Secretary Xi Jinping made a judgment that "development is still the key to solve all problems in China", and put forward that the requirements of development in the new period are "scientific development following economic laws, sustainable development following natural laws and inclusive development following social laws",[326] which are highly consistent with the sustainable development advocated by the international community, and set up correct development ideas for China to deal with the contradictions among economic, environmental and social interests.

In terms of many shortcomings caused by China's extensive and extensive growth mode in the past, General Secretary Xi Jinping clearly pointed out that "development is the Party's top priority in governing and rejuvenating the country, and we must persist in taking economic development as the central task and promote political, cultural, social and environmental progress and other aspects in a coordinated way",[327] He also emphasized that only development will deliver progress, and promoting economic progress is not only an important foundation for achieving Two Centenary Goals, but also an important foundation in pursing national prosperity, social stability and people's well-being; On the other hand, Xi stressed that development must be scientific and it must respect economic laws. The essence of scientific development is to maintain the balance, coordination and sustainability of development. He pointed out that the way out for economic development lies in improving economic quality and efficiency, and it should not be at the cost of the environment.

In view of the global ecological environment crisis, General Secretary Xi Jinping put forward: "All members of the international community should work together to build a sound global eco-environment. We should respect nature, follow nature's ways and protect nature. We should firmly pursue green, low-carbon, circular, and sustainable development",[328] so that green, circular and low-carbon development can be an effective way to transform the development mode and solve the ecological environment problems.

In terms of people's livelihood in development, General Secretary Xi Jinping pointed out that improving people's livelihood is the fundamental purpose and value of economic development, "the people's aspiration for a better life is our goal".[329] We must promote comprehensive, coordinated and sustainable development by putting people first and coordinate the efforts to stabilize growth, readjust the structure, promote reform and improve people's livelihoods.

Above-mentioned development concepts will not only serve as a guide for China's economic and social development, but also serve as a "China's answer" to enrich the theoretical system of global sustainable development and vigorously promote the global sustainable development process with China's practical actions. China participates extensively in global governance and shoulders the international responsibility and responsibility of building a community with a shared future for mankind. On the one hand, China gathers international strengths and resolutely builds international dialogue mechanisms and promotes international cooperation in the fields of climate change and afforestation; On the other hand, China gradually leads global governance as a leader.

I: Strengthening South-South cooperation and reshaping the new global governance order.

South-South cooperation is an important part of global cooperation and an important way for developing countries to unite for self-improvement, solidarity and mutual assistance and achieve common development.[330] China strengthens South-South cooperation, upholds the basic position of developing countries, and makes the vast number of developing countries an important force in the global governance structure, which is of vital importance for establishing the say over international regimes of global governance and reshaping the new global governance order including fairness, justice and common development.

In the past, the prominent theme of South-South cooperation was to reduce poverty and promote mutual economic and trade exchanges. Climate change needs to be shouldered by the whole world, so climate governance is increasingly becoming a major issue in South-South cooperation. When General Secretary Xi Jinping attended and presided over the Roundtable on South-South Cooperation at the United Nations Headquarters in New York, he put forward four suggestions for South-South cooperation in the new period: explore diversified development paths, better synergize development strategies, achieve practical development results and improve the global development framework, among which "expanding communication and exchanges with developed countries, establishing diversified partnerships, and building a community of interests of all parties" pointed out the path for South-South cooperation to promote multipolarization of global governance structure.

In recent years, China has provided strong support to the emissions reduction and development of developing countries in terms of capital, technology and projects. At the United Nations Summit on Sustainable Development, President Xi Jinping announced: "China supports developing countries, especially the least developed countries, landlocked developing countries and small island developing countries, to cope with the challenges of climate change. In September 2015, China South-South Cooperation Fund on Climate Change of RMB 20 billion was established and started in 2016. We will carry out 10 low-carbon demonstration zones, 100 climate change mitigation and adaptation projects and 1000 cooperation projects for climate change training in developing countries, and continue to promote international cooperation in terms of clean energy, disaster prevention and mitigation, ecological protection, climate-adapted agriculture, and low carbon and smart city construction and other fields, and help them improve their financing capabilities."[331] The multilateral financial mechanism is the main channel to promote the process of coping with climate change. "The South-South Cooperation Fund" is a financial mechanism independently established by China and cooperating with developing countries in coping with climate change. As a supplement to the Green Climate Fund, it can help developing countries to raise funds from the Green Climate Fund and other financial institutions or international institutions. It is an important tool for China to participate in global climate governance and will also become an important way for developing countries to achieve common development.[332]

II: Holding the G20 Hangzhou Summit and opening up a new era of global sustainable development.

As the host of the G20 Hangzhou Summit, China promoted all parties to jointly implement the 2030 Agenda for Sustainable Development, Addis Ababa Action Agenda and the Paris Agreement, and discussed major issues such as climate change, clean energy and green finance. General Secretary Xi Jinping pointed out: "Facing the current challenges, we should implement the 2030 Agenda for Sustainable Development and promote inclusive development. We put development at the top of the agenda of the G20 and jointly worked out an action plan to implement the 2030 Agenda for Sustainable Development. At the same time, we will reduce global development inequality and imbalance by supporting industrialization in Africa and the least developed countries, improving energy accessibility, developing Inclusive Financing, and encouraging youth entrepreneurship, so that people of all countries can share the outcomes of world economic growth."[333] China's leadership in the global sustainable governance system through the G20 platform fully demonstrates China's responsibility to lead the world move towards inclusive and coordinated development, and will create a new era of global economic growth and sustainable development, which is embodied in the following aspects.

Taking climate change as a priority, we call on member countries to sign and implement the Paris Agreement. Prior to the G20 summit, China and the United States ratified the Paris Agreement, and deposited the ratification instrument of the Paris Agreement on Climate Change between China and the United States with the UN Secretary-General in Hangzhou, so as to promote the entry into force of the Agreement in 2016. At the summit, G20 finance ministers and central bank governors called on G20 member countries to sign the Paris Agreement as soon as possible and promote its early entry into force, confirming that developed countries have fulfilled their commitments under the framework of the United Nations Climate Change Convention, including providing funds for developing countries to carry out mitigation and adaptation actions according to the agreement. At the same time, the G20 stressed that "taking urgent actions to deal with climate change and its impacts" proposed in the 2030 Agenda for Sustainable Development should be a priority, and encouraged to provide and mobilize more resources to deal with climate change, and encourage capital to flow to low greenhouse gas emissions and climate resilient development. It called on multilateral development banks and development financing institutions to incorporate actions to deal with climate change into their development strategies, and encouraged multilateral development banks to submit action plans to deal with climate change. These measures not only play an exemplary role in promoting the entry into force of the agreement, but also provide solutions to the financial and technical problems related to climate change, and promote substantial progress in global climate governance.

Incorporating green finance into development issues and promoting green, circular and low-carbon development. At the Summit, it is pointed out that in order to support global development under the premise of environmental sustainability, sustainable development should be regarded as the key field and specialized field of investment and financing, and financial services should be provided for improving the environment, coping with climate change and saving and efficient use of resources through "green finance". At the domestic policy level, the People's Bank of China, the Ministry of Finance and other seven departments and ministries have jointly issued the Guiding Opinions on Building a Green Financial System, taking the lead in making a demonstration.[334] The G20 Green Finance Research Group initiated by China submitted the first G20 Green Finance Synthesis Report to Hangzhou Summit and the voluntary optional measures initiated by it, which is conducive to enhancing the ability of the financial system to mobilize private capital to carry out green investment. In addition, the G20 also put forward measures to promote green finance development, such as providing clear strategic policy signals and frameworks, promoting the voluntary principle of green finance, expanding the learning network of capacity building, supporting the development of local green bond market, carrying out international cooperation to promote cross-border green bond investment, encouraging and promoting knowledge sharing in the field of environmental and financial risks, and improving the evaluation methods of green financial activities and their impacts. As the summit communiqué demonstrates, green finance provides an effective financial policy tool for sustainable development, and will also become an important guarantee and driving force for promoting the global green, low-carbon and circular development.

Putting forward specific action plans around sustainable development and implementing the 2030 Agenda for Sustainable Development. The G20 Action Plan for Implementing the 2030 Agenda for Sustainable Development formulated at the G20 Hangzhou Summit ensured the connection of the development strategy of this Summit with the 2030 Agenda for Sustainable Development. The high-level principle is put forward in the action plan to launch initiatives at the highest level in the world through its appeal, and then promote collective actions, including implementing the sustainable development goals and the Addis Ababa Action Agenda for Financing for Development, recognizing the importance of promoting the three major areas of sustainable development (economy, environment and society) in a comprehensive, balanced and coordinated manner, and realizing the importance of people-centered sustainable development. As a 15-year "dynamic document" formulated according to the sustainable development agenda, the action plan covers sustainable development areas such as promoting sustainable agriculture and rural development, sustainable infrastructure investment, energy accessibility, clean energy and energy efficiency, etc. It will be updated and adjusted according to the initiatives put forward by the future G20 presidency and the emerging needs, experiences and challenges, providing action guidelines for the implementation of the 2030 Sustainable Development Goals.

Spreading the advanced experience of structural reform and improving environmental sustainability. Under the new normal of China's economy, the structural reform with "addressing overcapacity, reducing inventory and deleveraging" as the core has been promoted. With the innovation-driven strategy, it plays an important role in maintaining stable economic growth and cultivating new impetus for economic development. Sharing structural reform at the G20 Hangzhou Summit contributed "China's answer" for leading the recovery of the world economy and ensuring environmental sustainability. The G20 Agenda for Deepening Structural Reform adopted at the Summit provided high-level guidance to all members. At the same time, it is advocated that the selection and design of structural reform should conform to the economic conditions of countries, especially defining "improving environmental sustainability" as one of the nine priority areas of structural reform, and incorporating environmental factor into the internal factors that promote growth, integrating with the development of agriculture, industry, energy and infrastructure to ensure the coordination of economic, social and environmental development.

Conclusion: Writing a New Chapter for Building an Ecological Civilization and Building a Beautiful and New World Featuring Harmony Between Human and Nature

Developing a Systematic Understanding of Theoretical Characteristics of Xi Jinping's Thought on Ecological Civilization

Nature is an interdependent and interactive system. As an integral part of the world universe, nature itself is also a whole, and its internal elements are interdependent and influence each other. As a system of objective existence, the existence and development of nature is independent of human will. Nature exists before human society. It can be said that human society comes after nature, which is the foundation of human society's existence and development.

On the one hand, Xi Jinping's thought on ecological civilization acknowledged the objective existence of nature as a whole. In terms of nature as an interactive system and the world as a unified system, General Secretary Xi Jinping profoundly pointed out: "The country's mountains, rivers, forests, land and lakes form a community of shared life. The geographical environment is holistic."[335] On the other hand, it has grasped the objective law of interdependence and mutual influence within nature, and in dealing with the relationship between man and nature, it is required to respect, conform to and protect nature first. Xi holds that the unified and interrelated world system and its development are regular. Xi Jinping's thought on ecological civilization keenly touches the essence and internal operation law of ecological civilization construction. He said that "The rise or fall of a civilization is closely tied to its relationship with nature". The success or failure of ecological environmental protection depends on the economic structure and economic development mode. "We must handle the relationship between economic development and ecological environment protection well." We should also persist in protecting while developing and develop while protecting, so as to realize the coordination of economic and social development with population, resources and environment.

Therefore, Xi Jinping's thought on ecological civilization itself is the natural dialectics of the relationship between man and nature, and it is the basic theory and method to guide people to correctly handle the relationship between man and

J. Pan, *China's Global Vision for Ecological Civilization*,
https://doi.org/10.1007/978-981-16-4534-1

nature, man and man, and man and society. General Secretary Xi Jinping pointed out more profoundly that "As for ecological environment protection, we must not cross the line, otherwise we should be punished"[336]. To understand and grasp Xi Jinping's thought on ecological civilization, we must deeply understand its theoretical characteristics from the perspective of dialectics of nature and philosophy of the relationship between man and nature.

(I) Problem orientation and bottom-line thinking

General Secretary Xi Jinping pointed out: "The Chinese Communist Party has always been engaged in revolution, construction and reform with the aim of solving China's real problems. It can be said that reform is driven by problems and deepened by solving them."[337].

The process of human beings' understanding and transformation of the world is a process of discovering and solving problems. Comrade Mao Zedong pointed out that problems are the contradictions of things, and where there are contradictions remained not to be solved, there are problems. The development of practice is never-ending, and the movement of contradictions is never-ending; new problems will arise when old ones are solved. Problems are the voice of the times, and every era always has its own problems. Only by establishing a strong sense of problems can we treat them pragmatically and find the roadmap that will lead us to progress. Xi Jinping's thought on ecological civilization is permeated with strong problem consciousness, clear problem orientation, firm belief and responsibility.

"(The Communist Party of China) All comrades must be clear about the urgency and difficulty of protecting the ecological environment and combating environmental pollution, as well as the importance and necessity of strengthening the ecological progress, and determined to control environmental pollution and build a sound ecological environment to create a good working and living environment for the people" [338] "Since the beginning of this year, China's smoggy weather and some serious pollution problems such as safety of drinking water in some areas and the high level of heavy metals in the soil have been exposed to the public, causing a strong social reaction."[339] "The 18th National Congress of the CPC emphasizes the need to achieve scientific development and to accelerate the transformation of the mode of economic development. If there is still extensive development, even if the goal of doubling the gross domestic product is achieved, what will be the situation with regard to pollution? I'm afraid the resources and environment will be completely overwhelmed by then."[340].

We must adhere to a problem-oriented approach, apply bottom-line thinking and draw an ecological red line. General Secretary Xi Jinping pointed out, "We must think about the worst-case scenario from the perspective of bottom-line thinking, strive for the best results and be more prepared and advantageous against trouble." [341].

The bottom-line thinking is a kind of positive thinking ability that objectively sets the lowest goal, stands at the lowest point and strives for the maximum expectation. At present, on the one hand, as pointed out in the report of the 19th National Congress of the CPC the construction of an ecological civilization has achieved notable progress.

The entire Party and the whole country have become more purposeful and active in pursuing green development, and there has been a clear shift away from the tendency to neglect ecological and environmental protection. Efforts to develop a system for building an ecological civilization have been accelerated; the system of functional zoning has been steadily improved; and progress has been made in piloting the national park system. Across-the-board efforts to conserve resources have seen encouraging progress; the intensity of energy and resource consumption has been significantly reduced. Smooth progress has been made in major ecological conservation and restoration projects; and forest coverage has been increased. Ecological and environmental governance has been significantly strengthened, leading to marked improvements in the environment. On the other hand, the 19th National Congress of the CPC has made new major strategic judgments on the new historical position of China's development. With decades of hard work, socialism with Chinese characteristics has crossed the threshold into a new era. This new era will be an era of building on past successes to further advance our cause, and of continuing in a new historical context to strive for the success of socialism with Chinese characteristics. It will be an era of securing a decisive victory in building a moderately prosperous society in all respects, and of moving on to all-out efforts to build a great modern socialist country. It will be an era whose contradiction is between unbalanced and inadequate development and the people's ever-growing needs for a better life.[342] China has seen the basic needs of over a billion people met, has basically made it possible for people to live decent lives, and will soon bring the building of a moderately prosperous society to a successful completion. The needs to be met for the people to live a better life are increasingly broad. Not only have their material and cultural needs grown; their demands for democracy, rule of law, fairness and justice, security, and a better environment are increasing. Building on continued efforts to sustain development, we must devote great energy to addressing development's imbalances and inadequacies, and push hard to improve the quality and effect of development. With this, we will be better placed to meet the ever-growing economic, political, cultural, social, and ecological needs of our people, and to promote well-rounded human development and all-round social progress. We must continue to intensify our efforts to resolutely curb the deteriorating trend of the ecological environment so that it can be gradually improved and optimized. Particularly urgent and realistic are the following issues. "The concept of the ecological red line must be firmly established. Our ecological problems have reached such a serious level that the most severe measures must be taken, otherwise not only will it be difficult to fundamentally reverse the general trend of ecological degradation, but the other ecological development goals we envisage will also be difficult to achieve."[343]In this context, the term "ecological red line" highlights the value and significance of the bottom-line thinking of General Secretary Xi Jinping in the construction of ecological civilization. Red lines for ecology is the bottom line, is the prerequisite, the bottom line and the basis for maintaining "resilience" in dealing with resource, environmental and ecological issues.

(II) The people's position and the drive for reform

A good ecological environment is the fairest public asset and the most inclusive livelihood benefit. General Secretary Xi Jinping pointed that the fundamental purpose of our Party in leading the people to build a moderately prosperous society in an all-round way and in carrying out reform, opening-up and socialist modernization is "to continuously raise the people's material and cultural standard of living and promote the all-round development of human beings through the development of social productive forces". "The test of the effectiveness of all our work will ultimately depend on whether the people are truly benefited and whether their lives are truly improved, which is the essential requirement of adhering to the principle of running the Party for the public's good and exercising the state power for the benefit of people."[344].

We must be persistent in addressing both symptoms and root causes and starting with the most prominent issues that affect people's lives, making great efforts to solve the current outstanding problems, exploring the establishment of institutional mechanisms that will work for a long time and mobilize the enthusiasm of all parties, improving the quality of the environment, protecting people's health, and making the urban and rural environments more livable and people's lives better.

"We must take the advantage of anti-driving mechanism and go with the flow to make ecological progress prominent. This is where public opinion lies. The people are not dissatisfied with the GDP growth rate, but have more dissatisfaction with the bad ecological environment. We must make trade-offs, what do we want? The ecology is very important from the point of view of whether the people are satisfied or not and whether they agree or not."[345].

These statements closely combine the Party's purpose with the realistic expectation of the Chinese people for a good ecological environment and a beautiful vision for ecological civilization, and take the realization of human's free and all-round development as its goal, which embodies the perfect unity of scientificity, class nature and practicality. In the long-term practice of revolution, construction and reform, the Communist Party of China has formed and developed the mass line, which is the lifeblood and fundamental work line of the Party. We must remain committed to serving the people and relying on them in all endeavors and the principle of "from the people, to the people.

General Secretary Xi Jinping pointed out: "The reform and opening-up is a game-changing move in making China what it is today, and a game-changing move for us to achieve China's two centenary goals and its great national rejuvenation." [346]Currently, although China has made great efforts in ecological environment protection in recent years, the situation remains challenging and complex. The overall situation of resource exhaustion and environmental pollution has not been fundamentally changed; Ecological and environmental problems present many new characteristics, and the internal pressure faced by sustainable development has not been fundamentally alleviated; Economic growth is achieved at the expense of seizing and destroying the ecological environment, and the consumption of natural resources is increasing, resulting in more and more damage to the environment. This must be solved by deepening the reform of ecological civilization system. General Secretary Xi Jinping pointed out: "It is necessary to promote ecological environmental protection with systematic engineering." [347]It is necessary to adhere to the overall advancement, make overall plans for all aspects, levels and elements, and pay attention to promoting mutual promotion, benign interaction and coordination among all elements, so as to achieve balanced and coordinated development.

(III) Philosophical deposits and dialectics

Philosophy is the wisdom of mankind. General Secretary Xi Jinping pointed out: "Learning philosophy and using philosophy is a good tradition of the Party", and it is necessary to "master the scientific world outlook and methodology, better understand the laws, and promote work more dynamically."[348].

Firstly, Xi Jinping's thought on ecological civilization adheres to the basic social contradictions analysis method. The contradiction between production forces and the relations of production, and the economic base and the superstructure exist in all social forms, which define the nature and basic structure of society and promote the development of human society from the low level to the high level. Secondly, it insists that material production is the foundation of social life. Productivity is the most active and revolutionary factor to promote social progress. The basic task in building socialism, the basic task is to release and develop the productive forces, and material production is the decisive factor of social and historical development. Today, China still insists on the important strategic judgment that development is still the key to solving all problems in China, and insists on the focused endeavor in national development and concentrated on construction, so as to push forward the continuous development of China's social productive forces and promote the unification of the continuous enrichment of things and the all-round development of people. Thirdly, it insists that the people are the creators of history. Socialism with Chinese characteristics is a cause for the people in their hundreds of millions, which is an endeavor for making life better for the people. We must stand on the people's stand, trust the masses, rely on the masses to promote development, respect the people's dominant position, respect the people's pioneering spirit, and draw wisdom and strength from the great creation of the people.

Dialectics: The open mind doesn't hinder the breadth of the world. All the theoretical features of Xi Jinping's thought on ecological civilization highlight the "problem orientation", "bottom-line thinking" and scientific thinking methods such as holistic perspective and system theory. In the final analysis, it is the concentrated expression of dialectical materialism thinking. In terms of "problem orientation", ha said: "Reform, which was first upon us by problems, goes deeper, in addressing the problems", "We must have a strong sense of problems, take major issues as the guide, grasp major issues and key issues for further research and thinking, find out the answers, and strive to solve a series of outstanding contradictions and problems facing China's development."[349]. In terms of "bottom-line thinking", Xi said: "To make decisions and do things, we must be good at using the method of bottom-line thinking, prepare everything from the bad situation, strive for the best results, be prepared, do not panic, and firmly grasp the initiative."[350]In terms of holistic perspective, he said: "Those who do not seek the overall situation cannot seek a domain. Everyone comes from different departments and units, so we should look at the problem from the holistic perspective. First of all, we should think whether the proposed major reform measures meet the overall needs and are conducive to the long-term development of the cause of the Party and the state. We should actually look ahead, think ahead and plan ahead. Only in this way can the final document truly meet the development requirements of the cause of the Party and the people."[351] As for the system theory, he said: "The reform of economic, political, cultural, social and ecological civilization and the reform of Party building are closely linked and integrated with each other. The reform in any one field will affect other fields, and it also needs close cooperation with the reform in other fields. If reforms in various fields are not matched, and reform measures in all aspects are related to each other, it will be difficult to deepen overall reform. Even if it is barely promoted, the effect will be greatly reduced."[352].

Developing an Overall Understanding of Historical Vision and Construction Essence of China's Ecological Progress

The 19th National Congress of the Communist Party of China comprehensively analyzed the international and domestic environments and China's development conditions, and put forward two phased goals for socialism with Chinese characteristics from 2020 to the middle of the twenty-first century, including the phased goal and historical mission of ecological civilization construction. In the first stage from 2020 to 2035, we will build on the foundation created by the moderately prosperous society with a further 15 years of hard work to see that socialist modernization is basically realized. The vision is that by the end of this stage, the following goals will have been met: There is a fundamental improvement in the environment; The goals of building a beautiful China is basically attained. In the second stage from 2035 to the middle of the twenty-first century, we will, building on having basically achieved modernization, work hard for 15 years and develop China into a great modern socialist country that is prosperous, strong, democratic, culturally advanced, harmonious, and beautiful.

In today's international context, China's peaceful growth is an irreversible historical trend. Industrial civilization has caused a series of resource problems, environmental disasters and ecosystem degradation problems. Industrialization, which follows the idea that man will conquer nature, has pushed the earth's energy and environmental crisis to a critical state. Under such circumstances, if the rising China continues to follow the road of industrial civilization, whether it is out of the need of China's own development or out of consideration of the evolution of human civilization, it can neither achieve harmonious, scientific and green development, nor realize the beautiful Chinese dream of the great rejuvenation of the Chinese nation. China must walk out of a new civilization model that avoids the disaster of human civilization and is suitable for human sharing.

In the 21st century, Xi Jinping's thought on ecological civilization has created the theoretical system and practical mode of ecological civilization construction, which is of great significance in the history of human civilization. As a result, socialism with Chinese characteristics has not only undergone the transformation of production mode, growth mode and development mode that is developed from the traditional extensive industry mode, but also self-reflection and adjustment of ecological ethics, moral consciousness and behavior mode, which requires the establishment of ecological civilization concept in the whole society and the formation of sustained, healthy and green consumption mode and great change of life style. This road, in essence, is developed from traditional industry civilization. Although ecological civilization is put forward based on solving the contradiction between human beings and ecological nature, the essential connotation of ecological civilization cannot be equated to protecting ecological environment. The essential connotation of ecological civilization is to create a new balance and coordinated relationship between human beings and nature, consumption and production, material and spirit, state and state, politics and culture under the guidance of the new civilization concept, and to build a new civilization shared by all mankind with ecological, intelligent and low energy.

China's Ecological Progress Promotes the Building of a Beautiful World and a Community with a Shared Future for Mankind

Human beings share an ecological home on earth. Resources and environment are the common wealth of mankind. We should promote the formation of a community with a shared future for mankind and interests. Ecological civilization is a philosophical issue essentially, which is the whole world outlook, cosmology, values and methodology of the harmonious development of man and nature and social and natural in the whole human society.

China's stance, world vision and human mind, and working together to build a win–win, fair and reasonable climate change governance mechanism have always been the global governance concept and major ecological concept of General Secretary Xi Jinping's continuous thinking, exploration and promotion of building a community with a shared future for mankind. [353]Since the 18th National Congress of the Communist Party of China, the thought of "a community with a shared future" has become a global concept with a higher vision and a wider space–time actively promoted by General Secretary Xi Jinping with global vision and human mind for governing the country. General Secretary Xi Jinping pointed out: "The international community is increasingly becoming a community of common interest. Some countries are getting richer, while others are poor and backward for so long. This situation is unsustainable. When a river rises, the boat becomes taller. When small rivers have water, the big ones are filled. When countries seek their own development, they should actively promote the common development of other countries, so that the development results will deliver more benefits to our people of all countries."[354].

As for the relationship between ecological civilization and the world, the ecological civilization construction initiated, vigorously promoted and practiced by the CPC and the Chinese government has been widely recognized by the world. The old practice of industry civilization is unsustainable. Both developed industrialized countries and developing countries that have not yet completed industrialization have realized that they need to abandon—or transform and upgrade with ecological civilization—the ethical value cognition, production mode, consumption mode and corresponding institutional mechanisms under industrial civilization. However, China's ecological civilization construction just provides systematic theory, method and policy experience, and has the ancient oriental ecological wisdom of Chinese traditional civilization. China has achieved the best results in implementing the Millennium Goals of the United Nations and made the greatest contribution to the world. The United Nations 2030 Agenda for Sustainable Development and the Paris Agreement, which were reached in 2015 and came into effect in 2016, are actually the agendas to promote the transformation from industrial civilization to ecological civilization. The harmonious concept of ecological civilization, the concept of harmony between man and nature, and the measures of social harmony and win–win cooperation can be seen everywhere in these two texts, and have become the main theme. The 5p (people, planet, prosperity, peace, partnership) concept of "2030 Agenda for Sustainable Development" embodies the integration thought of people-centered, earth protection, economic development, social harmony and win–win cooperation, which is similar to the thought of the Five-sphere Integrated Plan-China's overall plan for building socialism with Chinese characteristics including ecological civilization construction.

In terms of the great practice of building a community with a shared future for mankind, in order to actively participate in international cooperation on climate change, during the APEC in Beijing in 2014, China and the United States jointly announced that they would jointly introduce policies to reduce carbon emissions. The United States set the goal of reducing carbon dioxide emissions by 26% in 2025 compared with 2005. China announced that its carbon dioxide emissions will peak around 2030 and strive to achieve it as soon as possible. In 2016, the 2030 Agenda for Sustainable Development of the United Nations and the Paris Agreement to actively respond to global climate change were launched or officially entered into force. Paris Agreement is the first continuously effective and legally binding agreement adopted by all the parties in history, which is directly related to the major emitters including China actively taking responsibility and making efforts to reach an agreement. The Paris Agreement itself also provides external institutional constraints for China to accelerate the construction of ecological civilization, requiring the country to better fulfill its independent contribution plan and innovate the road of green and low-carbon development and energy transformation. General Secretary Xi Jinping thought deeply about the relationship between China's development and the world, and pointed out that in response to major global threats and challenges, developing countries should play a corresponding role to incorporate climate change into economic and social development plans, and vigorously promote green growth. He said that "China will continue to assume its due international obligations, carry out in-depth exchanges and cooperation with other countries in the field of ecological civilization, promote the sharing of achievements, and work together to build a beautiful home with good ecology."[355].

From the development trend of green industry in the world, a new round of industrial revolution and economic revolution with green technology and green industry as the core is sweeping the world. Major developed countries and emerging economies in the world are drawing up their own green development plans, hoping to win the first opportunity and seize the initiative in the world green development tide and gain the right to speak in green development. Therefore, green prosperity is the common dream of the whole world. China is committed to the construction of domestic ecological civilization, while actively promoting global sustainable development with the image of a responsible big country. With the enhancement of China's national strength and persistent efforts to promote green development, With the enhancement of China's national strength and persistent efforts to promote green development, China has gradually assumed the leading role in global ecological governance, contributing wisdom and strength in constructing regional cooperation system, maintaining global ecological security and leading global climate governance, aiming at building a community with a shared future for mankind and accelerating the process of global sustainable development.

Xi Jinping's thought on ecological civilization is inclusive, open and broad, focusing on coordinating the two major situations of international and domestic, world and nation, global and regional, jointly turning pressure into power, turning crisis into vitality, and jointly leading civilization inclusiveness, mutual learning and mutual acceptance. From the process of cooperating to cope with the global ecological crisis, we will strengthen cooperation with major developed countries and related developing countries in the new round of science, technology, industry and energy technology revolution and actively participate in the global environment to jointly build a clean and beautiful world.

Postscript

The eco-environmental field of Xi Jinping Thought on Socialism with Chinese Characteristics for a New Era embodies Chinese wisdom and program to promote the sustainable development of mankind and global eco-environmental construction. Since the industrial revolution, industrial civilization, which is characterized by relying on technology, giving priority to utility, accumulating wealth, and remaking and conquering nature, has rapidly ruled the world. While creating unprecedented material wealth, it has also led to increasingly prominent problems such as environmental pollution, climate warming, resource depletion, ecological degradation, growing gap between the rich and the poor, and a lack of social justice, threatening the survival and development of human society. Realizing the transformation of industrial civilization and seeking sustainable development have become the goals pursued by the world today. China's eco-environmental construction inherits Oriental way of thinking which holds that "heaven and man are united as one" and that "Dao operates naturally", and upholds the value concept that "lucid waters and lush mountains are invaluable assets". It attaches importance to the systematic thinking that mountain, river, forest, field, lake and grass land are related to each other and forms a Chinese-style eco-environmental development, transforming and promoting industrial civilization.

Xi Jinping's ideological system on eco-environmental construction adheres to the concept of respecting nature, following its laws and protecting it, and insists that protecting eco-environment means protecting productivity and improving eco-environment is increasing productivity. It encourages low-carbon, green and circular ways of development, which embodies the high integration and dialectical unity of the frontier theory and practical exploration of eco-environmental construction. Xi Jinping's ideological system on eco-environmental construction takes the wise ecological thinking of Chinese traditional culture as the soil of intellectual enquiries, the essence of Marxist dialectics of nature as the foundation of theoretical source, and the great exploration of eco-environmental crisis and construction in contemporary world and China as the practical source to select the essence and discard the gross of pure science.

J. Pan, *China's Global Vision for Ecological Civilization*,
https://doi.org/10.1007/978-981-16-4534-1

This is the whole content and overview to be revealed in this book concerning the eco-environmental field of Xi Jinping Thought on Socialism with Chinese Characteristics for a New Era. If this result is achieved, CASS Thinktank for Eco-civilization Studies, which the author's team relies on, will feel extremely gratified.

This book is mainly written by Pan Jiahua, who completes the organization and structure of the overall framework, the research and writing of part of the writing outline, preface and postscript, as well as the overall three times of final compilation and edit and revision. Zhuang Guiyang refines and enriches the writing outline, and is responsible for the research and writing of Chaps. 6, 8 and 9, Li Meng Chaps. 2 and 3, Lou Wei Chaps. 4 and 5, and Huang Chengliang Chaps. 1, 7 and 10 and the research and writing of part of the preface and postscript. Xue Supeng, Zhou Zhenge, Bo Fan, Shen Weiping and Wang Di contribute a lot in writing assistance and technical support.

Cai Fang listened to Pan Jiahua's report many times and put forward many constructive insights and specific guidance. Cheng Enfu put forward some suggestions on revising some theories and expressions in this book. Li Hongwei, Huan Qingzhi, Ouyang Zhiyun, Guo Zhaohui, Chen Ying, Wang Jingfu, Guo Zhanheng, Gu Yikang, etc. attended the expert demonstration meeting on the book and put forward valuable opinions on the revision and improvement of the manuscript. Zhao Jianying, president and editor-in-chief of China Social Sciences Press and Wang Yin, director of Large Publishing Series enthusiastically support the publication of this book and give specific help. I would like to express my heartfelt thanks to all of you.

Pan Jiahua

References

Marx and Engels' Classics and Comrade Xi Jinping's Main Works

1. Xi, J. (2007). *Zhejiang, China: A new vision for development*. Zhejiang People's Publishing House.
2. Xi, J. (2013). *Results-oriented and pioneering spirit: Reflections and practices for advancing Zhejiang's new development*. Central Party School Press.
3. Xi, J. (2014). *Xi Jinping: The governance of China*. Foreign Languages Press.
4. Xi, J. (2014). *Up and out of poverty*. Fujian People's Publishing House.
5. Xi, J. (2015). *Be a secretary of the county committee like Jiao Yulu*. Central Party Literature Press.
6. Xi, J. (2017). *Xi Jinping: The governance of China II*. Foreign Languages Press.
7. Party Literature Research Center of the CPC Central Committee. (Ed.) (2014). Excerpt of Xi Jinping's comprehensively deepen reform exposition. Central Party Literature Press.
8. Party Literature Research Center of the CPC Central Committee (Ed.) (2016). Excerpt of Xi Jinping's discourse on building a moderately prosperous society in all respects. Central Party Literature Press.
9. Party Literature Research Center of the CPC Central Committee (Ed.) (2017). Excerpt of Xi Jinping's exposition on ecological civilization construction. Central Party Literature Press.

Monographs

10. Fu, W. (2005). *From western philosophy to Zen Buddhism*. SDX Joint Publishing.
11. Liu, S. (2007). *Collected works of Liu Sihua on sustainable economy*. China Financial & Economic Publishing House.
12. Pan, J. (2015). *China's environmental governing and ecological civilization*. China Social Sciences Press.
13. Qiu, R. (1993). *Overseas philosophical issues in natural science studies*. China Social Sciences Press.
14. Shen, M. (2008). *Ecological economics*. China Environmental Science Press.
15. Wang, W. (2016). *The latest achievements of the sinicization of Marxism—A study of Xi Jinping's thought on governance*. China Social Sciences Press.
16. Zhuang, G. (2007). *How will China move towards a low carbon economy?* China Meteorological Press.

J. Pan, *China's Global Vision for Ecological Civilization*,
https://doi.org/10.1007/978-981-16-4534-1

17. Huang, C., & Yu, M. (2010). *Ecological civilization: A comprehensive transformation of human society*. Central Party School Press.
18. Commoner, B. (1997). *The closing circle: Nature, man and technology* (W. Hou, trans.). Jilin People's Publishing House.
19. Capra, F. (1989). *The turning point: Science, society and the rising culture* (Y Feng, trans.). China Renmin University Press.
20. Wolin, R. (2000). *The terms of cultural criticism* (G. Zhang, trans.). The Commercial Press.

Journals

21. Chen, H., & Pan, J. (2012). China's theoretical and practical progress in ecological civilization construction. *Journal of China University of Geosciences (Social Sciences Edition), 12*(5), 13–17.
22. Chen, Y. (2016). G20 gives new impetus to the implementation of the 2030 Agenda for Sustainable Development. *China Environment Supervision*, (8).
23. Dong, L., & Zhang, H. (2016). Environmental objectives in the 2030 agenda for sustainable development and its implications on the world and China in environmental Governance. *China Population, Resources and Environment*, (1).
24. Greene, H. F. (2006). Call for an ecologic society. *Studies in Dialectics of Nature*, (6).
25. Huang, C. (2015). Ushering in a new era of ecological progress under the guidance of four-pronged comprehensive strategy: Studying and implementing General Secretary Xi Jinping's important discourses on ecological progress. *Qiushi*, (8).
26. Li, F., Zhang, L., & Li, D. (2016). National park: New pattern for ecological and environmental protection in the three-river headwaters area. *Ecological Economy*, (1).
27. Li, M. (2015). The overall review of China's eco-compensation system construction in 2014. *Ecological Economy*, (12).
28. Liu, Z. (2016). China's contribution to global climate governance. *Qiushi*, (7).
29. Lou, W., & Pan, J. (2015). Discrimination of the concepts of "ecological red line" and "ecological bottom line". *People's Tribune*, (36).
30. Pan, J. (2013). Adapting to carrying capacity to ensure ecological security. *Social Sciences in China*, (5).
31. Pan, J. (2016). Construction, challenges and market expansion of emissions trading system. *China Population, Resources and Environment*. (8).
32. Pan, J., & Wang, M. (2014). Climate regime building in a changing world and China's role in global climate governance. *China Population, Resources and Environment*. (4).
33. Qian, X. (1994). Achieving the Sixth Industrial Revolution with modern science and technologies: Four letters about developing rural economy from Qian Xuesen. *Chinese Journal of Eco-Agriculture*, (3).
34. Sun, X., & Wang, L. et al. (2013). Insight into China's ecological civilization construction from a global vision. *China Population, Resources and Environment*, (7).
35. Tao, W. (2016). Developing a scientific understanding of Xi Jinping's thought on a community with a shared future. *Studies on Socialism with Chinese Characteristics*, (2).
36. Wang, R., & Zhao, Z. (2016). Green development building the ecological competitiveness of China. *Ecological Economy*, (2).
37. Xie, F., Cheng, H., & Li, A. (2014). An analysis of global climate control from the viewpoint of political economy. *Social Sciences in China*, (11).
38. Zhang, W., Jiang, H., Wang, J., Zeng, W., & Zhang, J. (2015). The role and contribution of science and technology innovation on the ecological civilization construction. *Chinese Journal of Environmental Management*, (3).
39. Zhou, H. (2015). Greening is the important part of modernization in China. *Chinese Journal of Environmental Management*, (3).

40. Zhu, F. (2012). Modern interpretation of Chinese traditional ecological culture. *Scientific Socialism*, (5).
41. Zhuang, G. (2014). The construction of ecological civilization system needs to seek breakthroughs in major areas. *Zhejiang Economy*, (14).
42. Zhuang, G. (2016). Responding to climate change and ecological progress under the new normal of economy: An interview with Zhuang Guiyang who is a researcher of the Chinese Academy of Social Sciences. *Yuejiang Academic Journal*, (1).
43. Zhuang, G., & Zhou, W. (2016). Model transformation of and China's contribution to global climate governance. *Contemporary World*, (1).

Newspapers

44. Huang, C. (2015). Ecological progress requires traditional ecological wisdom. *People's Daily*, p. 7.
45. Pan, J. (2015). Rethinking sustainable development economics. *People's Daily*, p. 22.
46. Zhou, G. (2016). Greening strategy of the Belt and Road construction. *China Environment News*, p. 2.
47. Zhuang, G. (2014). Solutions to high-carbon lock-in effect in urbanization. *Guangming Daily*, p. 8.

Index

J. Pan, *China's Global Vision for Ecological Civilization*,
https://doi.org/10.1007/978-981-16-4534-1

Printed by Printforce, the Netherlands